THE FALSIFIERS OF THE UNIVERSE

THE FALSIFIERS OF THE UNIVERSE

BIG BANG COSMOLOGY:
The first fraud in the final frontier

Bibhas De

Printed in the United States of America
Books by Bibhas De
First edition published: March 2015
www.bibhasde.com/thefalsifiersoftheuniverse.html

CONTENTS

PREFACE

This book is an outgrowth of my Internet campaign against the now nearly a century old scientific movement called Big Bang cosmology. This effort of mine started in the spring of 2007. At first my campaign was waged on my website *www.bibhasde.com*. It was then expanded to the blog site Dreamheron Chronicles (*dreamheron@wordpress.com*).

There are many reasons why I chose to wage this battle on the Internet rather than follow the conventional process of criticism practiced within the scientific establishment: submitting a critique to a refereed journal in the field. I only cite one reason here. Many dissident scientists, with far greater scientific stature than I, have tried for decades to follow this practice against Big Bang cosmology. They have not gained any ground. So I heeded the old advice about not doing the same thing over and over again and expecting a different outcome.

In another respect also my approach differs from that of the dissidents: their use of the polite language of academic discourse. The language of this book is the austere language of an investigative report, tinged at times with some sarcasm which I cannot resist when it comes to the Big Bang cosmologists.

Finally, I have no research background in cosmology and as such no bias for any particular cosmological theory. As an Arabic saying has it, I have neither a male camel nor a female camel in this matter.

1

THE BOOK OF THE ORIGINAL ORIGIN

The enormity of a false edifice is also a measure of the enormity of the task of tearing it down.
The history of development of Big Bang cosmology is presented in brief outline, by way of establishing the framework of the essential concepts.
It becomes clear that all the mathematics that has gone into constructing the framework would amount to something significant or to nothing at all depending only on whether or not the observed Cosmic Microwave Background radiation (CMB) could be shown to be the same as the predicted 3 K relic blackbody radiation.
When scientific logic is correctly applied, Hubble's law is shown to be an illusion of expansion of the Universe, and not a scientific discovery of this expansion.
Inflationary gravitational wave imprint pertains to the relic blackbody radiation and not to CMB, unless the two were shown to be one and the same.
Thus the answer to the question:
Is the Universe expanding, starting from an inflation era?
is the same in all respects as the answer to the question:
Has the 3 K relic blackbody spectrum been discovered?

> ... bizarreness masqueraded as creativity.
>
> Edward Gibbon, *The Decline and Fall of the Roman Empire*

CHAPTER I-1
Keyword: *Civilization*

The history of Big Bang cosmology is long – long in narrative and long in time span. Just as one cannot appreciate *The History of the Decline and Fall of the Roman Empire* without some appreciation of what it actually was that finally went down and keeled over, one must know a little about the growth of Big Bang cosmology to great heights to appreciate the various undercurrents that brought it to where it is today. The Roman Empire is actually an apt analogy, and we shall return to it.

Cosmology itself is a very old subject – perhaps as old as the Homo sapiens sapiens themselves. When the first of them looked up to the firmament and wondered, that was the beginning of cosmology. But we will stick to modern cosmology. It started as the bailiwick of the old-world variety savants in physics and mathematics and astronomy (and of course, philosophy) in the beginning of the twentieth century – more or less. Today in the space age we add to this the space scientists and the aerospace engineers and the computational scientists – the highest of the high-tech breed.

Today we speak only of *Big Bang cosmology* because that has been established by the powers that be as the true and the official account of the birth of the Universe. Other competing alternatives such as Steady State cosmology and Symmetric cosmology have now been sidelined or shelved.

Much has been written about Big Bang cosmology – in the scientific literature and then in the popular science literature and on to celluloid. Much contention has occurred between the scientific proponents and the opponents. The purpose of this book is not to add to these oeuvres. Nor does this book assail Big Bang cosmology on those grounds which the Big Bang opponents have already addressed. This book assails the accumulated experimental evidence that is said to have clinched the idea beyond any doubt, and so to have entered it into the realm of scientifically established facts.

Here is an indictment of the Big Bang cosmologists for their transgressions beyond the laxest boundaries of scientific ethics: First they scammed the world around Earth-based astronomical observations, and then they defrauded the world with space age technology in the high frontier. Then they were back doing the same thing again with Earth-based observations.

This process has become so perfected that they can now do their scamming blatantly and overtly. One group from among them performs some quack experiment with lots of high-tech bells and whistles backed by some hairy mathematics and some human interest stories, and the rest of them aver that all this is correct. Done and done. Yet another esoteric Big Bang idea becomes enshrined in this way in the pantheon of science. The larger scientific establishment approves or condones in silence.

Add to this the power base the Big Bang cosmologists have developed in the Government, the Legislature, and the media. It is now easy to see why they are unstoppable and untouchable.

By pretending to illuminate the civilization the Big Bang cosmologists have sullied it. If this book were to be assigned the most appropriate single-word bibliographic category to which it is most applicable, it would have to be:

Civilization

The sub-category would be:

Civilization, decline and fall

CHAPTER I-2
The Blessed Naissance

Monsignor Georges Lemaître, a Catholic priest and a professor of physics in Louvain, Belgium, liked to keep his faith and his physics separate. When Pope Pius XII issued an endorsement of his nascent scientific theory of origin of the Universe as a support for the biblical theory of creation, Lemaître reportedly was not happy.

That scientific theory, variously given the appellations *Single Quantum*, *Cosmic Egg*, *Atom Primitif* and *Primeval Atom*, holds that the Universe we see today started from the tiniest of tiny geometric point which suddenly exploded with tremendous energy. From within it came the stuff that eventually made our Universe. Reeling from that explosion, the Universe continues to expand to this day.

So this is indeed creation out of nothing. This scientific result of Lemaître happened to be – whether coincidentally or by design – the same as the biblical idea of creation.

By the end of the twentieth century, the expanding Universe idea achieved its crowning glory: space age technology clinched it good and proper. His Holiness Pope Benedict XVI now endorsed Big Bang cosmology as the official Catholic view of the Universe. The Catholic Church – it may be noted – commands fully one-sixth of the world population.

Now for a little backstory. Around about 1917, Albert Einstein applied his General Theory of Relativity to the Universe problem and found that there could not be a static Universe. The thing had to be either expanding or contracting.   To keep the thing from doing either of these and to make a static Universe possible, he introduced something called the Cosmological Constant in his theory (denoted by the symbol Λ.)

About the same time Russian cosmologist Alexander Friedmann performed an analysis of the same problem in the astronomical context and presented the case for accepting a non-static solution to the problem. All this so far, it should be noted, is very high mathematics, and nothing to do with anything real. This is the stage Lemaître stepped into.

 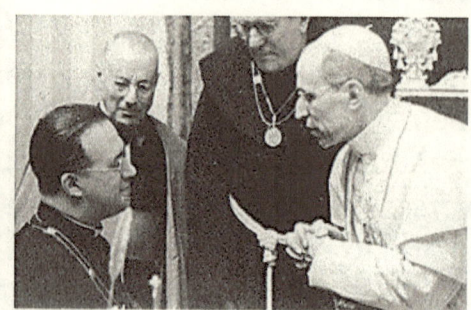

Left: Georges Lemaître and Albert Einstein.
Right: Pope Pius XII receiving Fr. Georges Lemaître.

Georges Lemaître was a remarkable multi-faceted man. He had impeccable academic credentials in mathematics and physics, and also in engineering. He had also studied philosophy. He did a stint of military service, receiving multiple decorations. All this he had under his belt before he turned to God. In sum, he was a practical man who was scientifically well-rounded and intellectually eclectic. Big Bang was born in no ordinary mind.

Upon publishing his idea on the expanding Universe in 1927 he became prominent in the scientific stage. He hooked up with Albert Einstein who initially did not like the idea. But Einstein gradually came around and became a mentor of Lemaître.

Around 1934 George Gamow, a Russian physicist and a

student of Friedmann, moved to America and settled down at the George Washington University. He quickly assumed a prominent position in the physics scene in general, and in the expanding Universe theory in particular. He was a theoretician, and worked in conjunction with his student Ralph Asher Alpher on how elements such as hydrogen and helium were synthesized in the early Universe. This contribution in nucleogenesis is considered in some quarters as a major load-bearing pillar of Big Bang cosmology.

wikipedia.org　　　wikipedia.org　　　　　　AIP

Alexander Friedmann, George Gamow, and Ralph Alpher

Clearly, the teacher-student trio Friedmann-Gamow-Alpher played a big role in the expanding Universe theory. George Gamow also took the ideas public – so to speak – through his flamboyant science-popularizing activities. He had a lively public confrontation going with British physicist Fed Hoyle, a prominent Big Bang dissident. This also served to keep the subject in the forefront of scientific news. Hoyle facetiously coined the name Big Bang for the primeval explosion, and it stuck. The expanding Universe theory acquired the permanent nomenclature: Big Bang theory and Big Bang cosmology.

George Gamow in fact became the first town crier for Big

Bang cosmology, and represented the beginning of a powerful thrust to propel the enterprise to great heights. This thrust occurred on two fronts: promoting the science and popularizing the science. In time, the mainstream science media would enthusiastically take over the role of popularizing Big Bang cosmology, and even deriding the rival cosmologies.

Saint Thomas Aquinas (1225-1274)

In time, an aura of eclectic intellectualism would be brought to the blessed naissance. Wise men saw an irresistible arena here for cogitating and agitating on the theological, spiritual and logician's dimensions of the dimensionless primeval atom. One thinkers' staple here was something called Thomistic Philosophy, after Saint Thomas Aquinas, a thirteenth century Italian Catholic priest and theologian. Proponents and opponents contended vigorously, talking volumes, talking Thomistic, talking pretty.

This sideshow would have a beneficial effect on preserving and protecting Big Bang. If the esoteric mathematical face of Big Bang scared away some would-be physics critics, the intellectual jabbering dissuaded some bread-and-butter physicists. Or else the history of Big Bang would be very brief indeed.

CHAPTER I-3
Linking theory to observation

In 1912 American astronomer Vesto Slipher first observed that a group of sharp spectral lines (Fraunhofer lines) in the spectrum of Andromeda Nebula was shifted to lower frequencies ("redder" frequencies) from where they were expected to be seen. He thus discovered cosmological redshift. In 1925 Swedish-American astronomer Gustaf Strömberg published more extensive redshift data on a number of galaxies. If this redshift were interpreted as Doppler shift, a shift in frequency that results from a velocity V of the source away from (redshift) or towards (blueshift) the observer, then a velocity could be assigned to the galaxies. The observed redshift thus meant that the galaxies are moving away from us.

About the same time two American astronomers Edwin Hubble and Milton Humason presented data on distances D to galaxies, based on their observations of stars in these galaxies and a set of assumptions.

These two groups of observational data would set the stage for transforming Big Bang cosmology from the highly mathematical abstraction that it then was to an observational science understandable by a much larger scientific community, and explainable to the public.

Lemaître had already introduced the theoretical concept of cosmological redshift in his work: the idea that light from a distant galaxy, say, would shift to lower frequencies in arriving to an observer on the Earth.

He explained this as a result of the stretching of space accompanying the expansion of the Universe. As light travels through this stretching space, its wavelength stretches the way a wiggle drawn on an inflating balloon stretches. The longer light travels through the stretching space, the greater is this lengthening of wavelength. So the farther away a galaxy is, the greater its observed redshift would be. Thus the concept of Doppler shift mechanism was not coopted into Big Bang cosmology. It became replaced by the stretching-of-space mechanism, today called cosmological redshift. Doppler shift was still retained in astronomy, and applied to nearby objects.

Clockwise from top left:
Vesto Slipher, Milton Humason, Gustaf Strömberg, and Edwin Hubble

From the above considerations, Lemaître was able to construct the relationship between the distance D to a galaxy and its expected redshift. He then compared this theoretical prediction with observational redshift data from Slipher and Strömberg, and the distance data from Hubble and Humason. The comparison was convincing (Figure I-1).

In 1927 Georges Lemaitre reported the value of what would later come to be known as the Hubble Constant. The values he reported were 670 km/sec/megaparsec and 575 km/sec/megaparsec, depending on how the data were grouped.

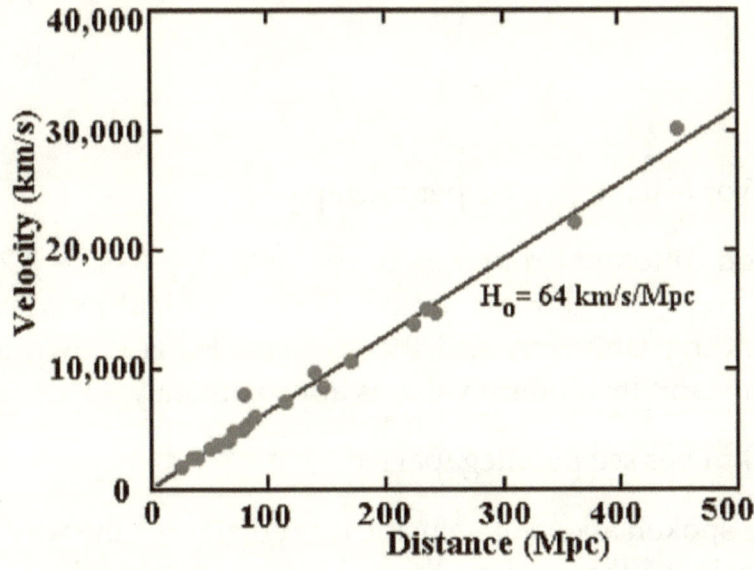

www.uni.edu/morgans

Hubble's Law (first reported in 1929) for distant galaxies. Here the velocities are derived from the redshift and the distances are derived by using supernovae as Standard Candles. The best value of the Hubble Constant H_o today is ~ 70 km/sec/megaparsec.

Figure I-1:
Hubble's law as presaged in the seminal paper of Georges Lemaître, and the Hubble Diagram.

13

The strip at the top of this figure is an image from Lemaître's original paper in French. The reader should also take in from this figure a sense of what the magnitudes of the cosmological expansion velocities are. As a reference, the velocity of light is 300,000 km/sec. The muzzle velocity of a rifle bullet is around a kilometer per second. Of course in Big Bang cosmology the objects are not colliding at their velocities shown and in this sense these velocities are not alarming.

If V is the expansion velocity of the galaxy, then Lemaître's comparison showed that there existed a relationship of the type

$$V = H_0\, D, \tag{I-1}$$

where

$H_0 \sim 575$ or 670 km per sec per megaparsec,

depending on different groupings of the data (1 parsec = 3.26 light years.) Today the above equation is known as Hubble's law of the expanding Universe, and the constant H_0 is known as Hubble's constant. Its modern value is approximately

$H_0 \sim 70$ km per sec per megaparsec.

We have spoken above of velocities and distances the way a motorist speaks of them. Here the velocity multiplied by the travel time gives us the distance between two towns. In actuality, these concepts are somewhat more involved in cosmology. For example, light takes a great deal of time to travel to us from a galaxy. By the time we see this light, that galaxy is no longer where it was when the light was emitted. So the distance we determine needs to be carefully interpreted. We note this fact here for completeness of discussion. In the following discussion this will be implied.

14

CHAPTER I-4
Cosmological distances

The observational confirmation of Big Bang cosmology rests crucially on the estimation of the distances D to cosmological objects such as galaxies and supernovae from the Earth. Today this estimation process is considered a completely settled issue, given especially the award of the Nobel Prize for Physics in 2011 for the discovery of the acceleration of the expansion of the Universe. That discovery rests very completely on regarding such estimations of distances as precision measurement science.

So how are these all-important cosmological distances determined? This is a very large subject and we want to touch only on the essential points. The first point we should note is what the word "determined" here truly means. It means "guessed" and not, for example, "measured". All cosmological distances are a guesswork. In no way does this come under the subject of mensuration.

To see this, let us start from scratch. A star has an optical *luminosity* L. It is the total optical power output (e.g., in Watts) of the star. If this star is at a distance D from us, then the amount of power incident on a unit area of a surface on the Earth perpendicular to the direction of the star is, by the inverse square law of spreading of radiation

$$\ell = L/4\pi D^2 \tag{I-2}$$

The symbol ℓ on the left hand side, called the *brightness*, is the thing we can measure with a telescope. The right hand side has two unknowns: L and D. We must provide one of these to find the other. There is no way we can do anything about knowing cosmological distances D. It is out of the question. Not even Star Trek style space-faring can get us this value. Nor can we *measure* the value of the absolute power output L of the star. This is the basic situation. The problem is not scientifically tractable unless we start to allow ourselves various types of leeway.

So the idea is to find ways to guess at the value of L. Once you have a guess value, you can plug it into the above equation and calculate a guess value of D. But the value of D will be as much a guesswork as the input value of L. That has been the basis of developing Big Bang cosmology. Right off the bat we have begun the guessing game.

And thus we come to the idea of *standard candles* in astronomy and cosmology. It is averred through assorted reasonings that the absolute luminosity L of certain stars throughout the Universe can be known with absolute certainty. These stars then become the standard candles, i.e., calibrator light sources with power outputs known to a high degree of precision. With yet another set of assumptions, unknown stars of measured brightness ℓ can be referred to these standard candles and hence the unknown luminosity L can be determined. Knowing L and ℓ, one can readily find the distance D from Equation (I-2).

There is really nothing scientifically wrong with proceeding this way into unknown territory when there is no other way to proceed. The problem arises when, over the course of time, this process of guessing is passed off as mensuration. And that is exactly how Big Bang cosmology has progressed: passing off wild guesswork numbers reported at one stage in time as, at a later stage, established precision numbers suitable for constructing discovery templates, referred to which newer and grander discoveries could be made and consummated.

In Big Bang cosmology, just the passage of time – just our mortal time – transforms assumptions and guesswork to clinched, hard science.

CHAPTER I-5
Cosmological velocities

I-5.1 Redshift: Stretching of space

Cosmological velocities, meaning here velocities of distant galaxies, supernovae etc., referred to the Earth, are determined from cosmological redshift measurements under the specific Big Bang assumption that this redshift is the result of stretching of space accompanying the expansion of the Universe.

It is useful to discuss this cosmological redshift alongside the conventional Doppler shift.

Doppler shift – a well-established phenomenon of physics – is a shift in frequency that occurs due to a relative motion between a source of radiation and an observer, along the line of sight. The original wavelength λ_1 (at emission at the source) of a spectral line in the radiation is shifted to a longer wavelength $\lambda_2 = \lambda_1 + \Delta\lambda$ when observed. The redshift parameter z is then defined by convention as

$$z = \Delta\lambda \, / \, \lambda_1. \tag{I-3}$$

The relationship between the redshift z and the v is

$$z = \frac{\sqrt{1+v/c}}{\sqrt{1-v/c}} - 1 \tag{I-4}$$

where c is the velocity of light.

In Big Bang cosmology a photon emitted from the source and traveling towards the observer travels through a space that is hypothesized to be continually stretching. The photon – essentially hugging this space – thus continually expands (the wavelength λ is stretched) during its flight, resulting in the redshift. So to find out the value of the wavelength at the end of the journey, we have to add up (by integration) the incremental redshifts suffered by the photon as it travels all the way from the source to the observer. This calculation leads to the relation

$$1 + z = \lambda_2/\lambda_1 = a_2/a_1 \qquad (I-5)$$

Here a is called the *scale factor*. The subscripts 1 and 2 refer to the time when the light left the source and when it arrived at the observer, respectively. A simple way to understand the scale factor for now is that if a_2/a_1 is 2.5, say, the Universe has expanded in linear size by two and a half times during the flight of the photon.

Generally speaking, Big Bang cosmology is concerned almost entirely with this space-stretching redshift. The Doppler shift in astronomical applications applies only to nearby objects.

There is an important point to note now. Unlike Doppler shift, the theory of cosmological redshift due to stretching of space has no support of any kind in observation. It is a highly esoteric mathematical construct. Space-stretching is not substantiated by the fact that the Universe is expanding (if in fact it is expanding). The discovery of redshift is not by itself an observational confirmation simultaneously of space-stretching *and* expansion of the Universe. If one of these is proved independently, then the other could be true.

I-5.2 Redshift: What is it in itself?

A tacit assumption made in Big Bang cosmology without any basis is this: The observed redshift is the result of some type of motion – whether it be the Doppler shift or the stretching of space. Alternative explanations of the observed redshift have been offered by others, and dismissed by the Big Bang cosmologists.

But let us step back from all this and ask the most rudimentary question.

What is the very first question a physics generalist, exercising nothing more than his physics gut, would ask when presented for the very first time with the puzzling data on redshift? I submit that the question is uniquely this:

Is there anything special about the cosmological situation with respect to the physics of light that we have not encountered before?

Then I submit that he would quickly answer his own question:

It is the extreme inverse-square dilution.

So he would end up with his final question:

What happens to the physics of light at extreme dilution?

We shall return to this question.

CHAPTER I-6
The nature of the Primeval Atom

According to Big Bang cosmology, if we take today's expanding Universe and extrapolate backward in time, then the Universe will get smaller and smaller, and theoretically, should collapse to a singular, dimensionless point: something with no length, no breadth, no height, and no volume.

So we come to the *geometric* idea that the Universe started from a single point. The beginnings of the entire Universe we see today – the galaxies, the stars, everything, even space itself – was packed within that point in some fashion. Outside the point, there was nothing – not even empty space. Before the point, time had not begun. Or perhaps we should say that the Big Bang clock had not been started.

This singular point was the Primeval Atom: the Cosmic Egg.

Now, something that already has a finite size – no matter how small – starting to expand is one thing. But going from a dimensionless point to something that has a size is another. How did *that* happen? Here is where the central idea of Big Bang came in. That point is not even a physical point. It is a mathematical singularity. It is said to have suddenly exploded with tremendous force (that is, tremendous energy) and to have become a real thing and to have created space as it expanded. But this explosion is actually not a conventional explosion. It is just an expansion that takes place with explosive, even supraluminal, speed. At the moment of the explosion the physics clock and the mathematics clock are started. This explosion came to be known as inflation and this birth of the Universe came to be known as Big Bang cosmology.

We cannot ask where this point was because there was no 'where'. We cannot ask what was there before this point, because there was no 'before'. We cannot ask what was outside the point, because there was no 'outside'. For the already expanding Universe, we cannot ask where the center of expansion is, for there is no center. Or every place is the center. Both statements work. We *can* ask when this was: Today the answer is said to be experimentally clinched. It was 13.798 ± 0.037 billion years ago today.

Things happened very quickly at the explosion – first in time scales of tiny fractions of a second (e.g., $\sim 10^{-43}$ sec – 10^{-34} sec.) Then things slowed down a bit – to timescales of minutes. First there was only energy in the growing Universe. Then matter started to appear: the sub-nuclear particles first appeared from out of this energy, in rough accordance with Einstein's mass-energy conversion principle. These particles then aggregated to form electrons, protons and neutrons. These then combined to form atoms, molecules, clouds of matter …. and on to galaxies and stars. When the Universe was merely about 300 000 years old, clumps of diffuse matter – the precursors to the present-day galaxies – appeared all over the place.

All the while this was happening, space was being continually created to accommodate all that was happening. Big Bang now unfolded as follows.

Most of the pure energy – the electromagnetic radiation that was released during the explosion – remained as electromagnetic energy. This energy, however, continued to interact (being absorbed, emitted, scattered) with the subatomic particles and thus the Universe remained opaque to this radiation.

Two things now happened to this radiative energy.

First, the stretching of space caused the wavelengths to become longer and longer as the expansion proceeded.

Second, as the subatomic particles began to combine to form

neutral atoms and molecules, the electromagnetic radiation became decoupled from matter. This means that the Universe became transparent to the radiation which thus became observable. This happened when the Universe was about 300 000 years old.

Modern Big Bang cosmology has it that this radiation – which expanded adiabatically – approached the well-known Planck spectrum. Planck spectrum is the spectrum of radiation emitted by idealized material – called *blackbodies* – which absorb every bit of radiation incident on them at any frequency, then assume an equilibrium temperature T and emit a radiation spectrum determined only by that temperature. Thus the Planck radiation and the Planck spectrum are also called blackbody radiation and blackbody spectrum (see Section III-2.)

So by the middle of the twentieth century it was beginning to be clear that two observations needed to be made in the sky in order to clinch Big Bang theory.

First and foremost, one needed to find in the sky that signature blackbody radiation spectrum – the relic radiation from the birth of the Universe. Because of the clean, pristine way this spectrum arose and evolved, it was expected to be observed as a textbook perfect curve – perhaps the most ideal example of a real-life blackbody radiation spectrum.

The quantitative discovery of the spectrum itself would provide primary support to Big Bang theory, and the qualitative discovery of a perfectness about the spectrum would provide additional support.

Second, when one found that radiation coming to us from that epoch, it should show us the image of the Universe as it existed then: A baby picture of the Universe (a more appropriately simile might be an ultrasound sonogram.) It should show us the clumping of matter, the protogalaxies. At a very fine scale the spatial map of the Universe seen *in the light of the relic radiation* should be patchy like cirrus clouds, and not homogeneous like an overcast sky.

This blackbody radiation is expected to be everywhere in the Universe and to be traveling in every direction. Also, the electric field in the radiation points every which way, meaning that it is unpolarized or randomly polarized. Another way of saying all of the above is that the relic blackbody radiation is *isotropic*. The said patchiness is a feature that is a very small perturbation over and above this isotropy. A discovery of this isotropy would be yet another point of support for Big Bang cosmology.

So now we can see that the observational clinching of Big Bang cosmology is the clinching of two predictions:

The Relic Blackbody Spectrum (primary prediction)
The Baby Universe Skymap (secondary prediction)

As to the temperature of this cosmic blackbody spectrum, Big Bang cosmologists have variously calculated this temperature, and the result has varied over time. The estimates ranged from a few degrees to about fifty degrees Kelvin. Ralph Alpher and Robert Herman – both students of George Gamow – estimated a temperature of about 5 K. Eventually, a number of about 3 K was settled upon. This is an extremely low temperature: 270 C below 0 C – the temperature of ice.

CHAPTER I-7
The mathematical cladding

Mathematicians and mathematical physicists were big players in Big Bang cosmology from the very early days. Alexander Friedmann and Georges Lemaître we have already encountered. In later times, one of the big players was (and is) the British physicist Stephen Hawking. Today there are many mathematical types laboring on Big Bang cosmology.

Famous among them are Alan Guth and Andrei Linde who specialize on the inflation era – the epoch starting at ~ 10^{-43} sec and ending at ~ 10^{-34} sec after the commencement of Big Bang.

wikipedia.org

Andrei Linde and Alan Guth.

What Guth, Linde and others did has now become a major chapter in Big Bang cosmology. More appropriately, this is a prelude to Big Bang theory, inflating the Universe from a point singularity to a golf ball or softball sized (say) affair. Big Bang then takes over this ball and expands it to the "present" Universe.

As a thinking aid, inflation theory can be seen as a customized adapter between mathematics Big Bang and physics Big Bang; between abstract concepts and real events.

In the inflation era, the volume of the Universe expanded by a factor of $\sim 10^{78}$ at supraluminal speed. Also during this period, the inhomogeneity of the Universe was seeded as a result of quantum fluctuations. These fluctuations at the subatomic length scales would later expand to represent the clumping of matter and thence to the formation of stars, galaxies etc. So the patchy Big Bang relic radiation skymap has its origin in the inflation era. After this rapid inflation, the nascent Universe would relax and grow more slowly.

In this way an entire elaborate chronology of the Big Bang Universe has been developed.

The main point to take from here is that all of this is of consequence because Big Bang has been proved to be correct by the observation of the 3 K cosmic blackbody radiation. If this were not so, then all the chronology and all the mathematics it is made of would remain as so much esoterica – interesting to some few and inconsequential to the rest of the world.

In 2014 Guth and Linde were awarded the Kavli Prize for Cosmology, along with the Russian inflation theorist Alexei Starobinsky. The award was justified by the fact that their ideas are so important that thousands of researchers today are working on them. The statement is most significant when one tries to think of how many other subfields of pure physics there are in which as many researchers work.

It is also important right here to identify one undercurrent in all this that is never discussed out loud: The issue of mathematics being used as armor. Most physicists are not mathematical wizards and they do not need to be. But if they try to criticize Big Bang theory, they are drawn into that arena where they may not feel quite comfortable. Then a conclusion is reached almost automatically that the critics are trying to assail what they do not understand.

CHAPTER I-8
Inflation era and gravitational waves

Something else happened in the inflation era. Gravitation arose and its energy was released in the form of gravitational waves (also referred to as gravitation waves.) Their corresponding signature today is expected to be found in a certain mode of polarization of the 3 K blackbody relic radiation.

Thus inflation era gravitational waves – if they are detected in this radiation – would offer strong support to the idea of cosmological inflation.

There is a trend among the Big Bang cosmology community to spread the word that inflation really is not a part of Big Bang cosmology. The purpose of this assertion is that if inflation should somehow be disproved, Big Bang should still survive intact.

This is not true. Both ideas require the existence of the blackbody relic radiation to be proved in the first place.

Before we move on, some basic concepts need to be introduced here for the purpose of our later discussion of inflationary gravitational waves. These concepts concern the polarization of the 3 K relic radiation.

As we will discuss later, in 1964 Arno Penzias and Robert Wilson discovered Cosmic Microwave Background radiation (CMB) in the Universe. This radiation was adopted by Big Bang cosmologists as their theoretically predicted 3 K blackbody relic radiation.

Generally speaking, CMB is observed to be isotropic and unpolarized. This means that the CMB intensity is the same in every propagation direction, and CMB electric field has no preferred orientation in the sky.

However, this is not quite true. We know that CMB has a

very faint patchy structure in the sky. And reportedly a very small portion of the CMB intensity is polarized.

For this polarized component of CMB intensity, the electric field of the electromagnetic wave has a specific directionality in the sky. That fractional intensity of CMB that is polarized and that directionality can be determined. Then a vector line segment can be drawn on the sky (as seen from the Earth) such that its length reflects the fractional intensity and its orientation reflects the orientation of the electric field in the sky. When such line segments are plotted for many adjacent observation points in the sky, we obtain a CMB polarization skymap.

Such expected skymaps can be divided into two categories, as shown in Figure I-2. The first category called E-mode polarization is where the bars form closed loops or radial spokes. The second category called B-mode polarization is where the bars forms twisted patterns and do not close the loops.

The E-mode polarization is more or less conventional polarization. It can be produced by such processes as scattering of CMB from free electrons in space.

If the distribution of matter in the sky were perfectly homogeneous, any polarizations arising through the above processes would of course cancel out, and no net polarization would be observed. It is the patchy, inhomogeneous structure that causes net observable polarization to arise. Thus there appear polarization swirls approximately outlining the patchy structures at the scale of observation.

The B-mode polarization is very special. It can be produced only by gravitational waves, which also produces E-mode signals. However, the B-mode signals are thought to be much weaker than the E-mode signals, which themselves are many orders of magnitude smaller than the CMB signal.

In recent times it has been argued that B-mode polarization can also arise from non-gravitational E-mode polarization through certain processes. This made the source of B-mode polarization swirls ambiguous – not unique to primordial gravitational wave.

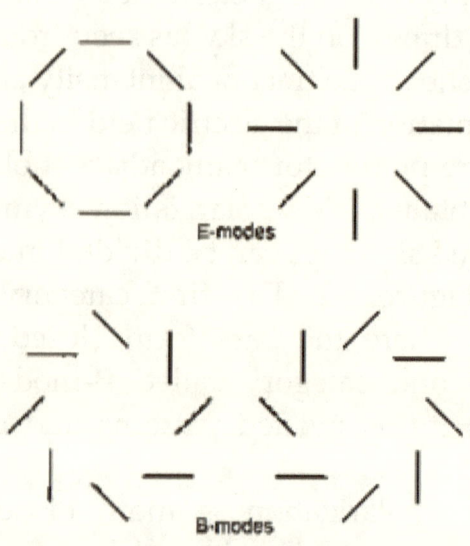

Figure I-2: E-mode and B-mode polarization swirls in the sky.

As an example, isotropic CMB radiation can be polarized by scattering from elongated dust grains in space. This E-mode polarization then can transform to B-mode polarization, and mimic the signature of gravitational waves.

So the point to remember is that the gravitational wave hunters would be looking for B-mode polarization patterns in the sky. If they find such patterns, they would then have to show that these did not arise from – wholly or substantially – a conversion of E-mode polarization from sources other than gravitational waves.

CHAPTER I-9
Dissidence

Initially, Big Bang cosmology developed in collegial contention with the other views on cosmology then prevailing. These ideas contended in the normal manner in which scientific ideas contend. The proponents of various cosmologies fought from nearly an equal footing. These cosmologies include Steady State Cosmology and Symmetric Cosmology. Steady State Cosmology, for example, accepts that the Universe is expanding but maintains that on the whole the Universe neither expands nor contracts.

Starting around the mid-1960s (when the Penzias-Wilson discovery of CMB was announced, and identified as the predicted 3 K relic radiation – see Chapter II-1), a triumphant Big Bang cosmology began to elbow out other views on cosmology. Gradually Big Bang cosmology assumed the role of the King of the Mountain and the others were consigned to lower grounds. Some of the proponents of these other cosmologies became the dissidents. This adversary role for them gradually overshadowed their creative role as the proponent of their own views of cosmology. An expressive name that has been applied to the dissidents is Big Bang Bashers.

As mentioned earlier the most visible Big Bang dissident was the British physicist Fred Hoyle, a proponent of Steady State Cosmology.

Another prominent dissident was Swedish physicist Hannes Alfvén, a proponent of Symmetric Cosmology. This idea says that the Universe contains equal amounts of matter and antimatter which are kept separated from each other by electromagnetic processes. Alfvén had paid his dues in mainstream physics and was by any reckoning a stalwart there.

He was a founding father of electromagnetic theory. He was probably the first scientist to examine Big Bang cosmology who was a well-rounded physicist. He was also an electrical engineer who designed and built with his own hands the electronic instrumentation (e.g., devices using the Rossi Coincidence Circuit) for his space experiments. Alfven had an eclectic learning in the classical sense. He also engaged in global societal issues of nuclear weapons ban and population control. In some ways Hannes Alfvén and George Lemaître were much alike.

wikipedia.org   wikipedia.org

Fred Hoyle and Hannes Alfvén

Alfvén's dissidence to Big Bang cosmology was not based so much on an attempt to preserve his own cosmology. It was rather that the Big Bang ideas did not agree with his physics gut. The very notion of a beginning (of time, of space, of gravitation) he did not like. Nor was he convinced that the playing of the video of the Universe backward would bring everything to a single point. Why would not the converging things bypass one another and expand again? Who can say that the expansion velocity vectors – when properly reversed – would be all convergent? Alfvén also did not like what he perceived to be a religious agenda behind Big Bang cosmology.

Hannes Alfvén was a strong proponent of the view (which he in fact founded) that electromagnetic phenomena were/are of paramount importance in many cosmological and astrophysical processes. Given that the Universe is filled with magnetic fields and plasmas in motion relative to these fields, he could not see how electromagnetic processes could be ignored altogether as it was done in Big Bang cosmology.

The report of the discovery of the 3 K cosmic blackbody spectrum in 1964 clinching Big Bang theory tripped up both Alfvén and Hoyle. They implicitly trusted that the discovery was legitimate, and each advanced his own alternative idea on how this blackbody came to be in the sky. But it is possible that this advent sapped them of their dissidence zeal somewhat. As we know today, there never was such a discovery (see Section VI-1.) The dissidents had been deceived into believing that the observational findings were legitimate scientific findings. Many of them passed away without ever learning about this deceit.

In the end Fred Hoyle and Hannes Alfvén were both sidelined in cosmology. The same fate befell the other dissidents such as Geoffrey Burbidge, Margaret Burbidge, Halton Arp, Jayant Narlikar, Eric Lerner *et al*. This list of dissidents is by no means inclusive. Countless other names can be added to this list.

Now a new breed of Big Bang verifiers would take over. While Penzias and Wilson observed one point in the Big Bang blackbody spectrum, this new breed wanted to define the full spectrum with more observational data points. Discovering the spectrum of the relic radiation from the birth of the Universe became the quest for the Holy Grail.

And it is easy to guess here that following any report of finding of this Holy Grail, the finding of the inflation era gravitational waves would be the next great quest.

CHAPTER I-10
Conclusions

Early Big Bang cosmology was developed and enshrined by a handful of workers. There was never any extensive community-wide scientific discourse or debate. There is no evidence that mainstream physicists were vigilant in evaluating new physics being introduced willy-nilly in Big Bang theory.

The interpretation of the observed redshift as due to stretching of space is an assumption that requires independent proof. Its introduction preempted the hardcore photon physics discussion of the redshift that needed to take place first and foremost.

The determination of cosmological distances using standard candles is subjective guesswork, not scientific mensuration.

Therefore the Hubble plot represents simply a transposing of the observed redshift-brightness diagram onto an assumption-guesswork (velocity-distance) diagram. This diagram is an illusion of an expanding Universe without containing any scientific discovery thereof.

The central prediction of Big Bang theory is the 3 K relic blackbody radiation. Once this radiation is found, a secondary prediction is the small-scale spatial anisotropy of this radiation. If the blackbody is not found, an anisotropy skymap found would have no relevance for Big Bang cosmology.

Inflationary B-mode polarization swirls are imprinted on the said 3 K blackbody radiation. If the blackbody is not found, any B-mode polarization swirls found in CMB would have no relevance for inflation theory.

If the blackbody radiation and the anisotropy are both observed, Big Bang cosmology could be posited for consideration as established science. The idea that the Universe is expanding might be considered well substantiated. And if the inflation era gravitational waves imprinted on this blackbody are also found, that might be considered the proverbial icing on the cake.

If the blackbody is not found, Big Bang theory is proved to be unverified.

If CMB is determined observationally to have a spectrum substantially different from a blackbody, Big Bang theory would stand solidly falsified in all its aspects.

BOOK II
THE BOOK OF DISCOVERERS

Those Magnificent Men in their Discovery Machines
might be a more imaginative title for BOOK II.
Over the past half a century our scientific civilization has
been blessed with the emergence of a new breed of
discoverers who guided us beyond the Firmament History
of the Book of Genesis,
and back into the Original Beginning.
Systematically, page by page, they extended the Genesis
backward in time, back 14 billion years ago today.
With phenomenal exactitude, they described for us what
happened in that epoch.
And in the process they ushered in the age of mind-
numbing scientific precision, beyond the wildest dreams
of our run-of-the-mill, everyday physics experimenters
toiling in the basement laboratories of physics
departments.
We were not ungrateful. We anointed and indulged these
new age explorers unstintingly; no holds barred.
We enshrined them permanently.

All men dream, but not equally. Those who dream by night in the dusty recesses of their minds, wake in the day to find that it was vanity: but the dreamers of the day are dangerous men, for they may act on their dreams with open eyes, to make them possible.

T. E. Lawrence (a.k.a. *Lawrence of Arabia*)

CHAPTER II-1
Penzias and Wilson

Arno Penzias immigrated to the United States from Germany when he was just a little boy, and his family settled in New York. He proceeded to receive his Ph. D. and then to join the AT&T Bell Laboratories in Holmdel, New Jersey. Here he met Robert Wilson, a born and bred Texan. They became collaborators in experiments having to do with antennas and microwave. Their company had some leftover equipment from a previous project in long distance communication, and the two researchers obtained permission to do some radio astronomical experiments with the equipment. The year was 1964.

Whatever Penzias and Wilson were thinking, cosmology was not on their minds. They did not even know about blackbody relic radiation in the sky proposed within Big Bang cosmology.

The antenna used by these researchers was a horn-like antenna, also called a sugarscoop antenna because of its shape (Figure II-1). It was a steerable antenna that could be pointed to various parts of the sky, down nearly to the horizon and even groundward. They operated their antenna near 4080 MHz, or a wavelength of about 7.35 cm. Their first task was to determine the antenna noise temperature, a parameter needed to establish the noise threshold of their system.

To their great bafflement, Penzias and Wilson measured an antenna temperature of ~ 6.7 K, well in excess of the expected ~ 3.2 K. This latter number was determined by taking into account all known sources of noise in the environment. They were now left with an unexplained excess antenna noise temperature

$$\Delta T_A \approx 3.5 \pm 1 \text{ K.}$$

They tried to troubleshoot this and pinpoint the source of this excess temperature. They did not succeed. The source of this temperature – indicative of an excess microwave power at 4080 MHz incident on the antenna – remained a mystery.

nps.gov

Figure II-1:
The Penzias-Wilson antenna in Holmdel, New Jersey.

Further studies with the antenna showed that this excess power was omnipresent and always present; and isotropic and non-polarized. These tests would have been done by pointing the antenna in various directions at various times of the day and the year, and by rotating the polarization plane of the antenna. An auxiliary antenna also might have been used.

～〜〜〜

Around 1960 Princeton physics professor Robert Dicke acquired a graduate student named James Peebles – a Canadian

38

who had immigrated to America. This would be the beginning of a beautiful relationship – especially as it concerned establishing Big Bang cosmology as the official history of the origin of the Universe.

Robert Dicke had distinguished himself as an engineering physicist, especially during his tenure at the famous MIT Radiation Laboratory, dedicated then to defense research. His name became famous in connection with the Dicke Radiometer – a type of microwave receiver. Dicke's work at the Laboratory had already caused him to think about measuring the predicted cosmic relic blackbody radiation, and so he was well situated to contribute to Big Bang cosmology – more so now that he had a very able graduate student in Peebles.

Dicke had in fact done an experiment from a rooftop of the MIT RadLab, and was able to establish that any relic radiation temperature T_{BB} would be no greater than ~ 20 K.

The teacher and the student now started in right earnest on nurturing and furthering the science of Big Bang cosmology. On a hot summer day in 1964, while discussing the possibility of there being present this relic blackbody radiation in the sky, Dicke told Peebles: "Why don't you go and think about the theory." And that set the lifelong scientific pursuit for Peebles. He was the theoretician, and looked after that side of the research. Dicke set out to try to observe the relic radiation.

~~~~~~

When Robert Dicke heard about the mysterious noise observed by Penzias and Wilson, he instantly connected this to the relic radiation that he himself was looking for – and was probably on his way to success. He might well have become crestfallen at being thus "scooped". But it seems that Dicke's devotion to his scientific pursuit was greater than his personal ambition. He saw in this development great opportunities for the cause. He and his group arranged a meeting at Princeton with Penzias and Wilson.

Arno Penzias, Robert Wilson, James Peebles, and Robert Dicke

We do not know what was discussed at that meeting between the Bell Labs group and the Princeton group, but we do know what followed. Two back-to-back papers in the same issue (July 1965) of *Astrophysical Journal* were published by the two groups. In effect, the first one – by the Princeton group – prepared the psychological ground for accepting the 3.5 K observation as the telltale Big Bang blackbody. The second paper – by the Bell Labs group – then detailed the observation. The two papers thus jointly pitched the discovery:

$$T_{BB} \approx \Delta T_A \approx 3.5 \, \text{K}.$$

This was the beginning of the era of Big Bang cosmology as hardcore science. After all, its very specific prediction of a relic blackbody in the sky that had developed over a long period of time had at last been confirmed in a spectacular way, and with a serendipitous discovery no less. Things snowballed from there. There was no stopping now, no looking back. And too bad for the dissidents! They did not have a leg to stand on anymore. For who can argue with such success?

The handy way of referring to the relic radiation today as "3 degree K blackbody" probably has its roots in this discovery. This number gave the Big Bang theoreticians for the first time a firm anchor point from which to venture out as far and as wide as they fancied. Whatever the significance of the discovery was perceived to be to Big Bang cosmology in 1964, it would only grow from there.

Outside of Big Bang cosmology and to science in general, however, the discovery – interesting though it was – was not of any great impact. A radio hiss, some television hash – these are the things it explained.

Most curiously, the plainly evident fact did not seem to bother anyone in the scientific community that a single frequency measurement near 4 GHz was being interpreted to confirm an entire blackbody spectrum that peaks out near 200 GHz, and is skewed to boot.

CHAPTER II-2
John Mather

The hunt for the 3 K cosmic blackbody had been on for quite some time before the idea of NASA's COBE satellite was conceived in the late nineteen seventies. Besides the serendipitous discovery of Penzias and Wilson, there had been other experiments specifically designed to look for the relic radiation in the sky. These experimenters realized early on that the best chance to observe this was to get the instrument package above the radio interferences in the environs and the atmosphere of the Earth. Thus the early experiments concerned balloon-borne and rocket-borne instrument packages. From these experiences would come the further conclusion that the best place to put the instrument package would be on an Earth-orbiting satellite.

In the early 1970s John Mather, Norman Nishioka and David Woody, doctoral students of Paul Richards, a physics professor at the University of California, Berkeley, were conducting balloon-borne experiments. The first flight of this instrument – which was to yield results for the doctoral dissertation of Mather – failed. However, he was awarded his Ph. D. in 1974 for his research work leading up to this experiment. A second flight of this instrument produced marginal hints of a blackbody spectrum, reported by the group in 1975 (Figure II-2). In 1981 Woody and Richards reported the measurement of a fuller spectrum – historically the first full spectrum measurement. This paper also had John Mather as a coauthor. The balloons in these experiments achieved heights of around 40 km.

In a couple of years' time after receiving his Ph. D., Mather ended up in NASA Goddard Space Flight Center where the idea of the COBE satellite was taking shape. The plan for the Cosmic

Background Explorer Satellite reportedly came about as a desire on the part of the NASA leadership to undertake a high impact scientific project. The idea evolved through various planning stages, ups and downs thereof, and reviews and reevaluations, to finally become a full-fledged program. Somewhere along the line, a certain strategy was adopted: COBE was to be an in-house NASA program, both science-wise and engineering-wise. What that meant was that unlike most other NASA projects where the actual engineering developments are done in contractor companies, in the case of COBE the design was to be done largely by NASA scientists and engineers.

The satellite was designated to be carried in the cargo bay of a Space Shuttle and deployed in orbit. It was being designed accordingly. However, after the 1986 Space Shuttle Challenger disaster the plan had to be abandoned. Eventually, the plan was revived and it was decided that the satellite would be launched on a Delta rocket. The size of the satellite had to be scaled down and consequently, substantial redesigning of the instrument package had to be done on short order.

The satellite was launched on 18 November 1989 from Vandenberg Air Force Base in California. It achieved a polar orbit about 900 km high (Figure II-3). This altitude was chosen as a compromise between getting far above the terrestrial interferences and staying far below the electromagnetic interferences of the Earth's radiation belt. The satellite orbited roughly along the day/night boundary, with an orbit period of 103 minutes and a spin rate of 0.8 revolution per minute. Its collecting apertures (or look directions) pointed away from both the Sun and the Earth. Its spin axis was along the look direction of Mather's Winston Cone antenna (*loc. cit.*). The scientific ground control for the satellite was located at NASA Goddard Space Flight Center in Greenbelt, Maryland.

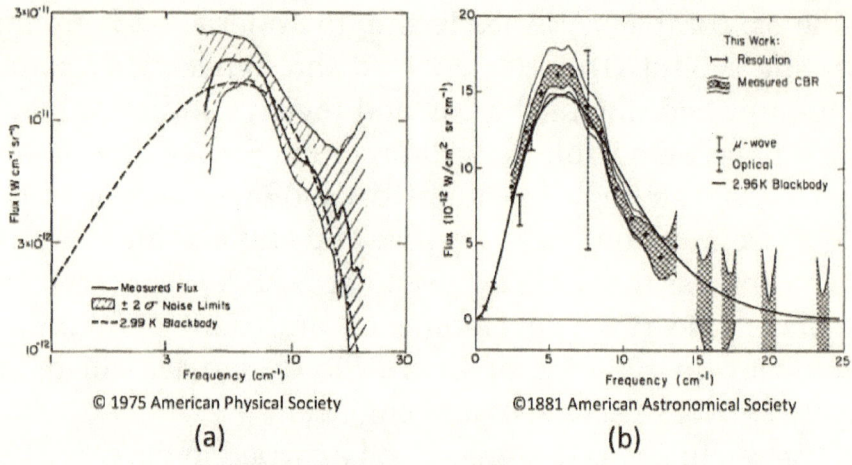

(a)

(b)

Figure II-2:
Balloon-borne full spectrum measurements: (a) Experiment of Woody, Mather, Nishioka and Richards (1975); (b) Experiment of Woody and Richards (1981).

berkeley.edu   NASA

Paul Richards and John Mather

There were three experiments on board the COBE satellite:

- *The Far Infrared Absolute Spectrometry (FIRAS)* experiment designed to measure the spectrum of the sky radiation in the approximate frequency range ~ 10 – 600 GHz. The objective was to clinch Big Bang cosmology once and for all by obtaining the full blackbody spectrum with dense data coverage for the very first time. (Principal Investigator: John C. Mather, NASA. In addition to being the scientific leader of the FIRAS experiment, Mather was also the overall scientific manager of the COBE Project.)

- *The Differential Microwave Radiometer (DMR)* experiment designed to generate the radiation skymap at frequencies of 31.5 GHz, 53 GHz and 90 GHz – near the peak of the expected relic blackbody spectrum. The objective was to detect small spatial anisotropy in the blackbody radiation (over and above its expected isotropy) – indicative of early clumping of matter to form the stars, galaxies etc. (Principal Investigator: George F. Smoot, University of California, Berkeley, and Lawrence Berkeley National Laboratory.)

- *The Diffuse Infrared Background Experiment (DIRBE)* designed to measure infrared sky brightness. The objective was to quantify infrared emission from cosmic dust. (Principal Investigator: Michael Hauser, NASA.)

The configuration of the antennas for these experiments are shown in Figure II-4. The collection of data with the satellite began a few days after launch. The electronics in the satellite were kept refrigerated by cryogenic liquid Helium at 1.4 K. The lifetime of the instrumentation was largely determined by the loss of this cooling. The FIRAS instrument ceased operating on September 21, 1990. The other instruments ceased operating on

December 23, 1993.

The primary problem of clinching Big Bang cosmology is the problem of making a high definition measurement of the entire predicted blackbody spectrum in the sky, and not just a point or two of the spectrum. This requires making measurements at sufficient number of close-packed frequencies so as to convincingly define an observed spectrum and its peak. There are basically three different approaches to this problem:

(1) Single broadband antenna, a frequency multiplexer and a multi-channel receiver.

(2) Multiple narrowband antennas with individual narrowband receivers.

(3) Single broadband antenna and a spectrometer.

In preparation for the design of the FIRAS experiment, the team settled upon Option (3) above (as with the previous Berkeley experiments), and for this purpose, designed and studied the properties of the Winston Cone (as with the Berkeley experiment) as a broadband radiation collector in the relevant frequency range. It is not known if any other options were considered in detail.

The choice of this option meant that the instrument would have the ability – for the very first time – to measure the blackbody spectrum from an orbit with nearly continuous data points. These points would unambiguously define the curve. There would not be any questions, any shadow of a doubt. The other options above would be limited to producing only a few data points to define the entire curve.

A parenthetical note: the previous Berkeley experiments under the guidance of Paul Richards, himself a Ph. D. from Berkeley, set the general philosophy of instrumentation that would continue in the NASA COBE satellite experiment a decade later: the use of the Winston Cone as a broadband collector of radiation, the use of interferometry for the full spectrum analysis of this radiation, and the use of synthetic blackbodies as calibrators. If NASA conducted an in-depth engineering review of this Berkeley experimental philosophy as imported to NASA by Mather, it did not result in any major changes.

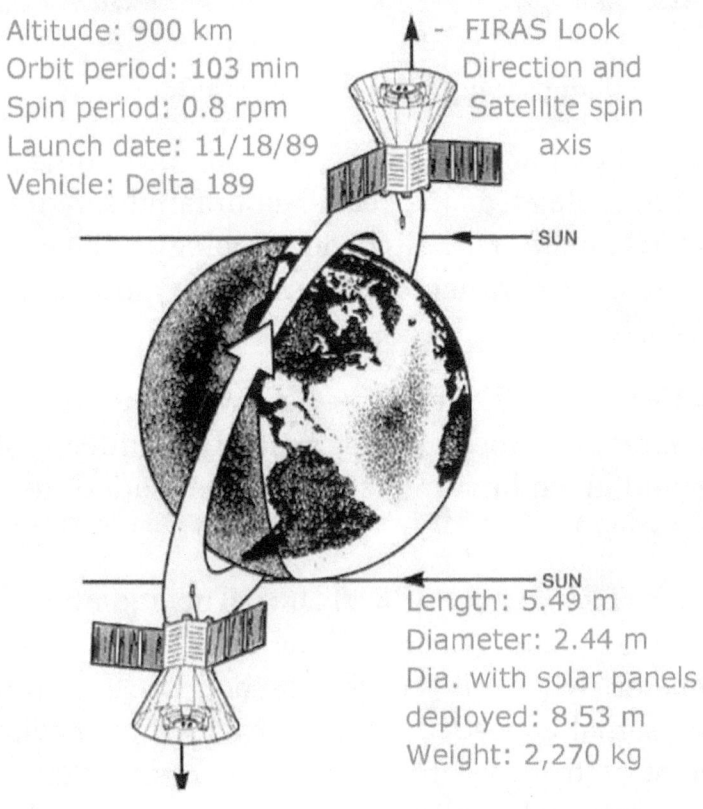

Altitude: 900 km
Orbit period: 103 min
Spin period: 0.8 rpm
Launch date: 11/18/89
Vehicle: Delta 189

- FIRAS Look
  Direction and
  Satellite spin
  axis

SUN

SUN
Length: 5.49 m
Diameter: 2.44 m
Dia. with solar panels
deployed: 8.53 m
Weight: 2,270 kg

NASA (modified)

Figure II-3: The orbit of NASA COBE satellite.

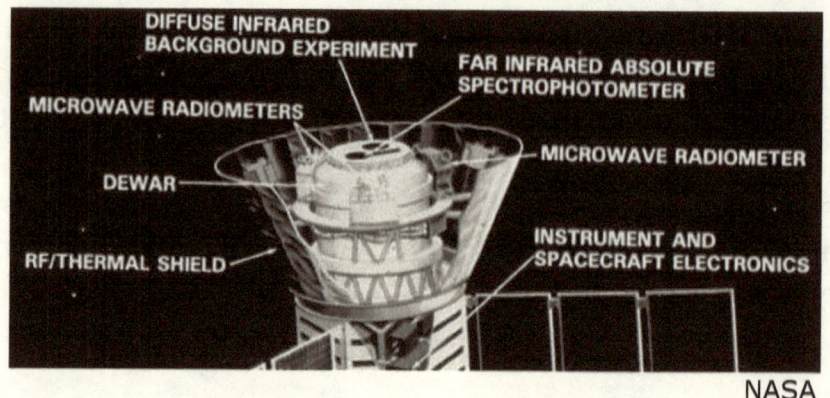

Figure II-4: COBE satellite configuration.

In the following discussion we will use some technical antenna terminology. If the reader is unfamiliar with this, no need to worry. In Book III all the terminology will be defined.

The FIRAS experiment had essentially four components (Figure II-5):

THE ANTENNA: The antenna here was a modified Winston Cone. A flared, trumpet-shaped section was added to the entry aperture and a collimating section was added to the exit aperture (*loc. cit.*).

THE EXTERNAL CALIBRATOR: The radiation received from the sky by the Winston Cone antenna is passed to an interferometer where it is processed. For this purpose a calibration signal is needed. This is provided by a calibrator in the shape of a "trumpet mute" made of a near-ideal synthetic blackbody material which emits a near-ideal blackbody spectrum at its temperature. It is mounted on hinges so that it normally stays clear of the Winston Cone antenna aperture but periodically swings in to snugly cap the aperture of the antenna. Thus the sky and the calibrator alternately illuminate the Winston Cone in the same way (same ray geometry inside the

48

cone) so that the two signals can be properly compared. The external calibrator is referred to as Xcal.

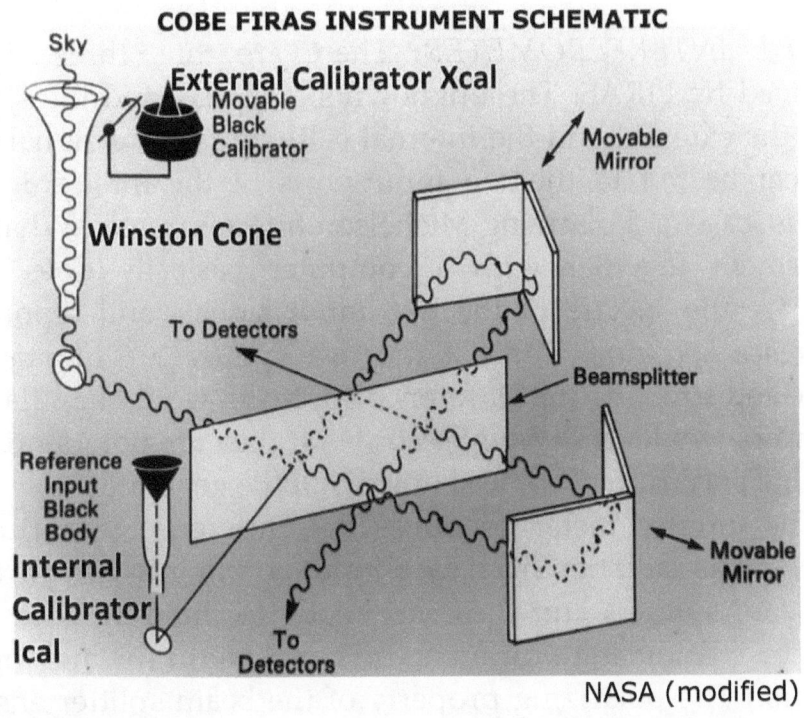

COBE FIRAS INSTRUMENT SCHEMATIC

NASA (modified)

Figure II-5: The FIRAS measurement technique.

THE INTERNAL CALIBRATOR/REFERENCE SIGNAL GENERATOR: Since the radiation received from the sky and that from the Xcal are not being received simultaneously, they cannot be fed to the interferometer simultaneously. For this reason a reference signal – generated internally in the satellite – is included as an intermediate step. Clearly, this reference signal has to be as broadband as the blackbody spectrum being sought. The reference signal generator is a smaller version of the main Winston Cone, and has its aperture capped permanently with a near-ideal blackbody trumpet mute. The reference signal generator is also referred to as the internal calibrator or Ical.

Both the internal and the external calibrators had arrangement for precision temperature measurement and control.

THE INTERFEROMETER: There are thus three signals observed by FIRAS: The unknown sky signal, and the signals from the external and the internal calibrators. Two signals at a time can be fed to the two input ports of the interferometer, in this case a Polarizing Michelson Interferometer. What this instrument together with a computer basically does is to compare the spectra of the two input signals and report the *Difference Spectrum*. Thus, if the two spectra are identical in shape and strength, the Difference Spectrum would be a flat line at zero power level (a *Null Spectrum*.) If they are not identical, a residual spectrum of some shape would be generated.

The interferometer produces the Difference Spectrum as follows: The radiation from each input is split into two parts (in the beam splitter) and these are made to interfere with each other by introducing a difference in path lengths (by moving the mirrors.) The polarizing property of the beam splitter ensures that, even though the plane of polarization of the input signals are not known, the comparison between the signal components takes place in the same plane of polarization. The signals are finally fed to detectors.

The mosaic of numbers thus generated from the detector output – the detector voltage vs. the mirror position, for example – can be processed (transformed) in the computer to generate the Difference Spectrum. It should be noted that the interferometer is dealing with the radiation at the electric field level. It is these fields that are combined (subtracted from one another.) The final detected output of the interferometer is in terms of power levels.

For the ease of discussion, we will use power levels rather than electric fields in the following.

The Winston Cone antenna was used for its extreme broadband nature. Its design parameters were chosen to fit the needs of FIRAS.

Figure II-6 shows the FIRAS antenna design and the design parameters. As mentioned previously, two main modifications were added to the basic Winston Cone. First, a fluted "trumpet bell" section *ABCHJK* was added to extend the aperture of the Winston Cone. Its purpose was to lower the sidelobe levels and thus improve the broadband characteristics. Second, an elliptical concentrator section *EDGF* was added to the backport of the Winston Cone. Its purpose was to collimate the concentrated radiation emerging from the said backport and to transform it to an approximately parallel beam (coming out of the backport of the concentrator.) The acceptance angle of the overall antenna was 7°. The signal from the backport was now incident on the input port of the interferometer.

This antenna stood at the scientific core of the experiment. It was the sample collection device – the most crucial component that is novel to this experiment. Everything else used in the FIRAS instrument was known and proven technology – which is not to say they were simple or easy.

The term sample collection I deliberately introduce here is of seminal importance. As a broad concept, it is as important here as the same term for a criminologist at a crime scene. If the sample collection is not done right by the criminologist and is not fully defensible, all the scientific quality in the subsequent laboratory analysis of the collected samples would be rendered completely irrelevant in a court of law. The same is true with the FIRAS experiment in the court of science.

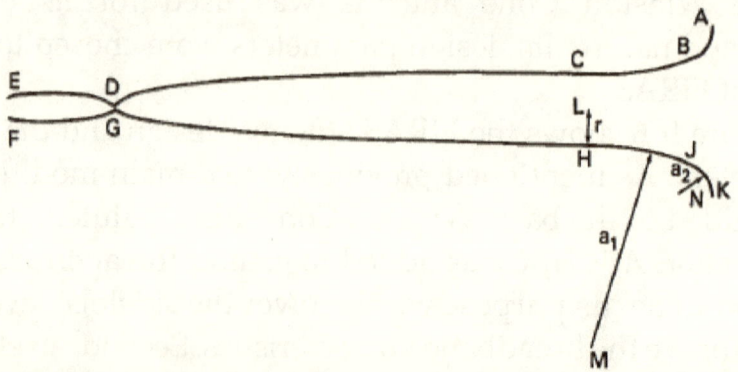

Fig. 1. Horn antenna consisting of compound parabolic concentrator *CDGH*, flare *ABCHJK*, and compound elliptic concentrator *EDGF*. Curve *CD* is a parabola with focus *G* and axis *GH*, while curve *GH* is its symmetric partner with focus *D* and axis *DC*.

Table I.   Dimensions of CPC and CEC

Diameters	*AK*	30.00	cm
	CH	13.65	cm
	DG	0.779	cm
	EF	3.98	cm
Lengths	*AK-CH*	20.86	cm
	CH-DG	95.36	cm
	DG-EF	19.79	cm
	EF-focus	25.00	cm
	HM	49.48	cm
	JN	6.04	cm
Angles	CPC acceptance	3.5°	
	Parabola axis tilt	3.5°	
	Large radius flare (*HMJ*)	20°	
	Small radius flare (*JNK*)	70°	

NASA

Figure II-6: The FIRAS Winston Cone antenna specifications.

The FIRAS experiment used the particular calibrator technique described above in order to circumvent the need of knowing the antenna patterns in detail over the frequency range of the expected blackbody spectrum.

For a period of time FIRAS antenna would collect the unknown radiation spectrum from the sky – presumably a blackbody radiation field (at the temperature T_{sky}, say.) Then the calibrator would swing in on a hinged arm and cap the aperture of the antenna. The inside volume of the antenna would now be filled with the blackbody radiation from this external calibrator at its temperature T_{xc}.

The sky radiation and the radiation from the external calibrator are sequentially compared with the radiation from the internal calibrator at a temperature T_{ic}.

We will use the notation S_v for the brightness of the spectrum of radiation in the sky and C_v for the brightness of the blackbody spectrum. When italicized, they will represent the corresponding power levels emerging from the backport of the antenna and impinging on the input port of the interferometer. These power levels have different units than their non-italicized counterparts. To be absolutely clear:

S_v: Sky brightness [Watts m^{-2} Hz^{-1} Ster^{-1}]

C_v: Brightness seen by the antenna capped by a blackbody [Watts m^{-2} Hz^{-1} Ster^{-1}]

S_v: Total spectral power level at the backport of the Winston Cone when it is looking at the sky [Watts Hz^{-1}]

C_v: Total spectral power level at the backport of the Winston Cone when it is looking at a blackbody [Watts Hz^{-1}]

Now to the measurement technique. This takes place in the

following steps (refer to Figure II-5):

When the FIRAS Antenna is looking at the sky, the interferometer output provides the difference in the two power spectra:

$$P_{sky,ic} = S_v(T_{sky}) \sim C_v(T_{ic})$$

When the FIRAS Antenna is next looking at the external calibrator, the interferometer provides the difference in the two spectra:

$$P_{xc,ic} = C_v(T_{xc}) \sim C_v(T_{ic})$$

Next, the difference between the above two differences can be obtained:

$$P_{sky,xc} = P_{sky,ic} \sim P_{xc,ic}$$

Now FIRAS obtains its final result. The temperature T_{xc} is adjusted until the two spectra on the right hand side above cancel each other – frequency for frequency – resulting in a Null Spectrum (zero at every frequency in the range of the blackbody spectrum of interest):

$$P_{sky,xc} = 0 \text{ at all } v.$$

When this happens,

$$T_{sky} \equiv T_{xc}.$$

So the ability of the FIRAS instrument to obtain a Null Spectrum in this way verifies two things:

(1) S_v is as perfect a blackbody as the external calibrator; and

(2) T_{xc} is the blackbody temperature of the sky radiation.

This is basically how the FIRAS results are obtained – at a conceptual level. But superimposed on this were enormous complications that we will discuss later.

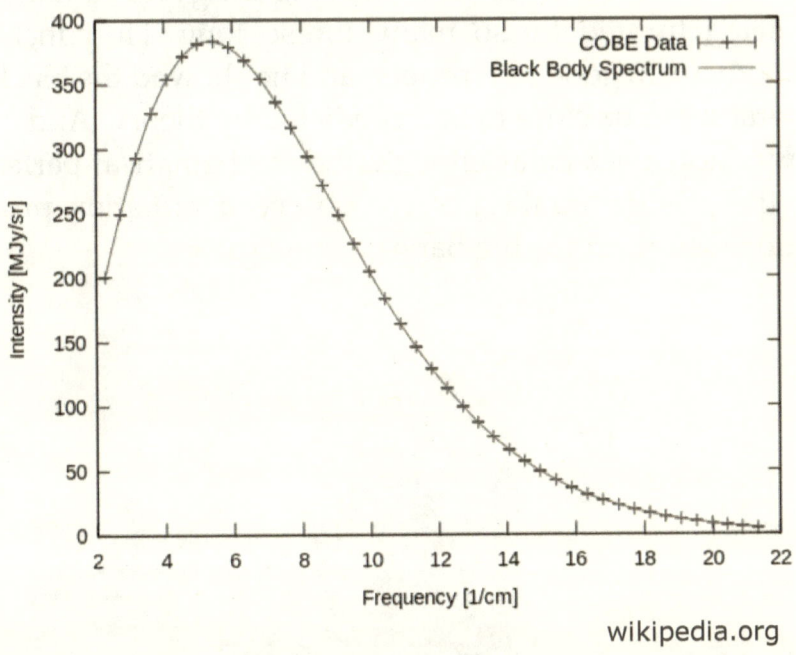

Figure II-7: The Big Bang blackbody relic radiation spectrum discovered by John Mather, compared with theoretical blackbody curve at 2.725 K.

In the end it was reported that the sky spectrum S_v is a nearly perfect mathematical blackbody spectrum at 2.725 ± 0.001 K (Figure II-7). It is so perfect that the errors bars on the data points were smaller than the thickness of the theory comparison line drawn. The errors were estimated to be of the order of 50 parts per million (of the peak intensity) over the major portion of the spectrum.

When these conclusions were presented for the first time at a meeting of the American Astronomical Society in the

Washington D.C. area in January 1990, John Mather received the now famous standing ovation, something that does not normally happen in such professional scientific meetings. Not only did Mather discover the telltale Big Bang relic radiation that was being hunted by so many for so long. He clinched its blackbody form good and proper, and he showed the blackbody temperature to be close to that predicted by theory. And as if all this was not spectacular enough, the mathematical perfectness (e.g., the pristine quality) of the observed radiation provided additional support for Big Bang cosmology.

> We have the job of building what has never been built before, to discover what was never known before.
>
> John Mather

CHAPTER II-3
George Smoot

John Mather's FIRAS experiment on NASA COBE satellite established beyond a shadow of doubt a certain frequency region of cosmic microwave background radiation as Big Bang cosmology's relic radiation. COBE satellite was also tasked to make a secondary investigation, the DMR experiment under the leadership of George Smoot, to verify if the sky seen *in this relic radiation* is isotropic or anisotropic. For both Mather and Smoot, a co-equal colleague in their projects was Charles Bennett.

Big Bang theory holds that the Universe seen in this light would show signatures of early clumping of matter. This would be a picture of the nascent Universe (i.e., the light DMR would observe was emitted in the epoch when the stars, galaxies etc. had just started to form.) However, the theory also holds that the departure from isotropy would be extremely small. The relic radiation is expected to be necessarily isotropic. There is then expected to be a small degree of anisotropy superimposed on it.

The interpretation of the Smoot experiment as a verification of Big Bang cosmology would thus be contingent upon the blackbody spectrum itself having been verified in the first place by Mather. As mentioned in the previous chapter, it was.

George Smoot in the late 1970s was a postdoctoral researcher at the Lawrence Berkeley National Laboratory, affiliated to the University of California at Berkeley. Like Mather, Smoot also had a background in trying to perform early versions of his COBE experiment, using other methods (most notably, flying instruments on a converted U-2 spy plane.)

Figure II-8 shows Smoot's DMR experimental set-up designed to map the Universe in the FIRAS frequency region. He chose to make his measurements at three single frequencies of the relic radiation: 31.5 GHz, 53 GHz, and 90 GHz. These frequencies are well within (near the peak in fact) the relic spectrum studied by Mather and at the same time have relatively small contribution from non-Big Bang cosmology sources such as galactic emission.

berkeley.edu   wikipedia.org

George Smoot and Charles Bennett

At each frequency, Smoot had a pair of identical antennas optimized for that frequency. These antennas were of the corrugated horn type, and all the antennas had the same beamwidth of about 7°, nearly the same as the FIRAS antenna beamwidth. Each antenna pair was mounted so that the two antennas looked at different patches of the sky, with identical beams. The angle between the look directions of the antennas was 60°. Thus each antenna was looking at an angle that is 30° off the spin axis of the satellite.

The radiation received through the two antennas were then combined at the microwave stage to produce the difference of the power received by the two antennas. Note again that it is the electric fields that are combined, although we loosely speak of power differences.

Thus, if the sky brightness was perfectly isotropic, the differential power observed from the two portions of the sky would be zero. If the sky brightness had a patchiness to it, a finite difference signal might be observed, depending on where the two antennas were pointed. The smaller the variation in the brightness, the smaller would be this difference signal. The differential power observed can now be assigned to the different portions of the sky.

When the antenna pair sweeps the entire sky, the accumulated data can be correlated to give us a contiguous map of such sky brightness.

Smoot had earlier been engaged in very similar experiments conducted from high-flying U-2 aircraft in the late 1970s. The angled antenna combination he used on the COBE experiment owes its design to that experiment. The corrugated horn design also came from there (Figure II-9). Thus, as with Mather, Smoot also imported his experimental philosophy from Berkeley to the NASA project.

The Smoot experiment turned out to take much longer than Mather's to analyze. He finally reported his results in 1992 in the form of skymaps of intensity variation (Figure II-10). Oval maps of this type show the celestial sphere in Mollweide projection (equal area projection.) For comparison, the map of the world is shown in the same projection.

The skymaps showed a distinct clumpiness of the Universe, or an anisotropy. The clumps have a range of angular size scales – as would become much clearer from later satellite skymaps (Figure II-11). Smoot's skymap provided a secondary confirmation of Big Bang theory, and according to some, a confirmation of the inflation theory. This map is one of the most iconic maps in science, showing us nothing less than a baby photo of the Universe from the Big Bang family album.

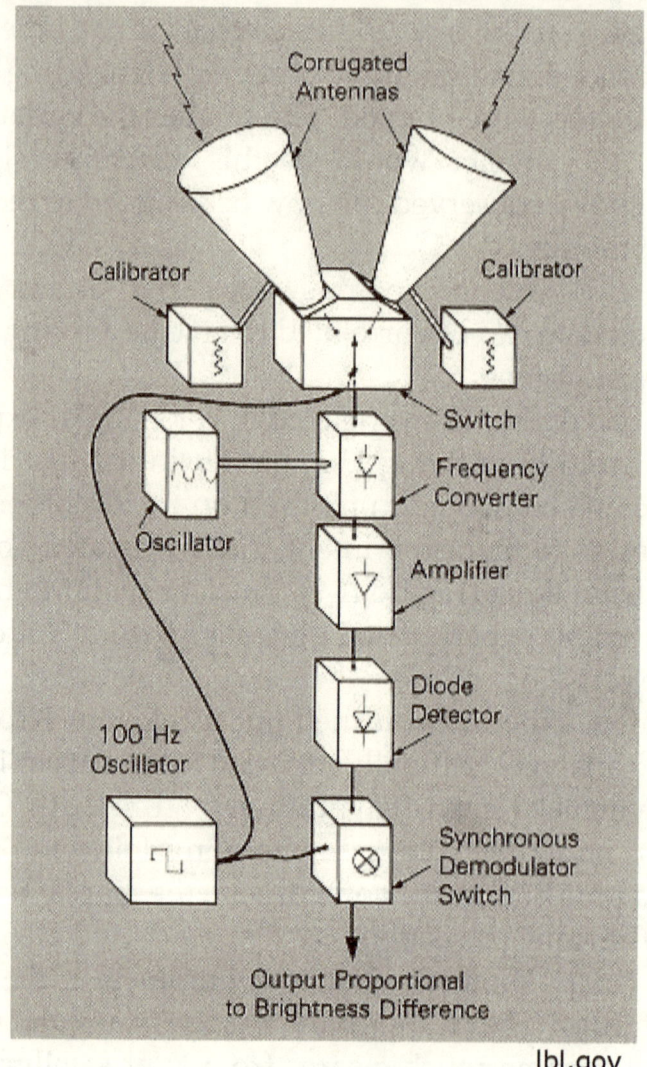

Corrugated
Antennas

Calibrator

Calibrator

Switch

Frequency
Converter

Oscillator

Amplifier

100 Hz
Oscillator

Diode
Detector

Synchronous
Demodulator
Switch

Output Proportional
to Brightness Difference

Figure II-8: The COBE DMR experiment:
(a) The measurement scheme.

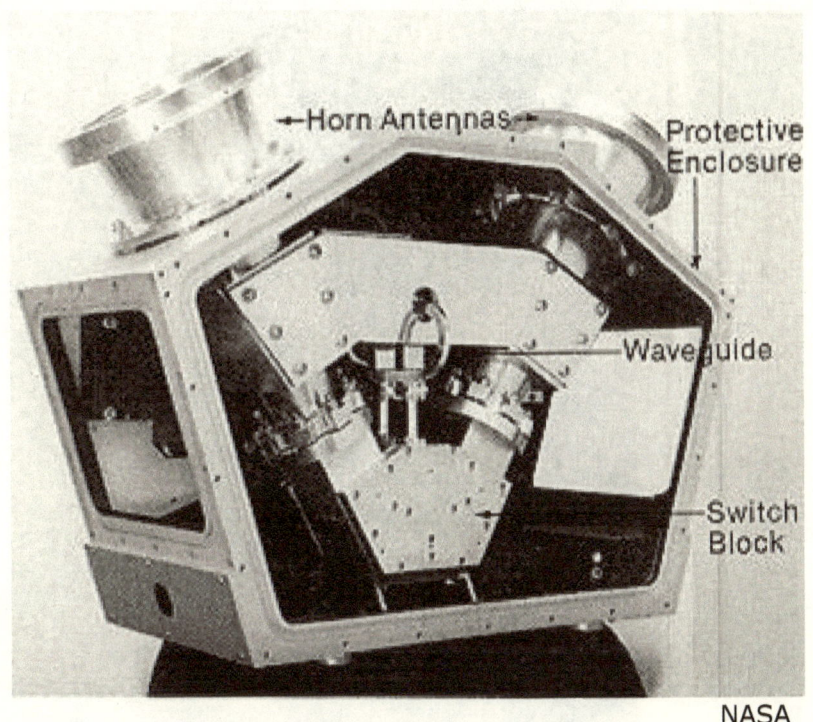

Horn Antennas

Protective Enclosure

Waveguide

Switch Block

NASA

Figure II-8: (b) One of Smoot's three pairs of corrugated horn antennas. The two antennas here look at different parts of the sky, separated by an angle of 60 degrees. The power received by the antennas are subtracted one from the other, thus generating a difference signal.

lbl.gov

Figure II-9: The design of the 34 GHz dual horns and their internal corrugation from Smoot's U2 experiment to determine microwave anisotropy of the cosmic background radiation.

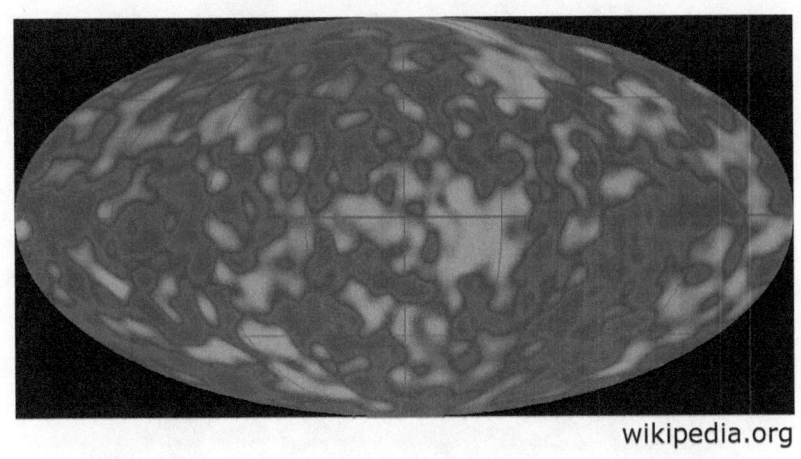

wikipedia.org

(a)

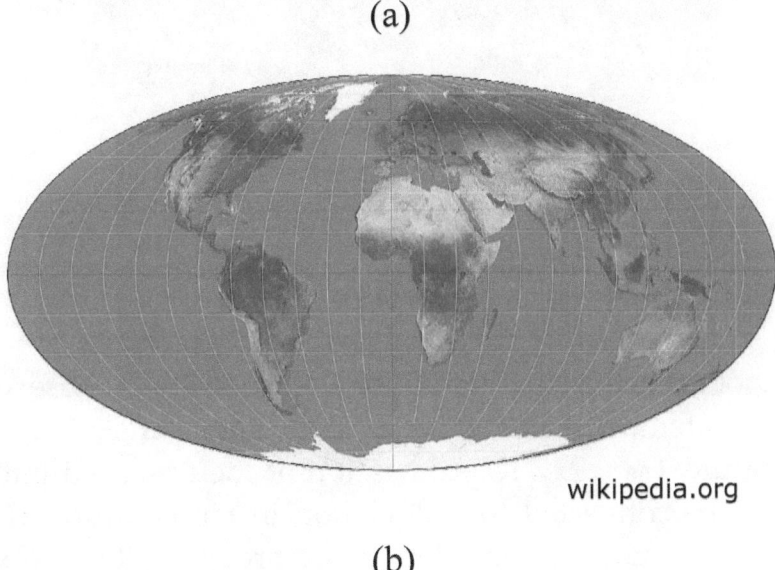

wikipedia.org

(b)

Figure II-10: (a) The Smoot discovery of the map of the early
Universe. The lighter regions are colder. (b) The World in Mollweide
projection (presented for clarity.)

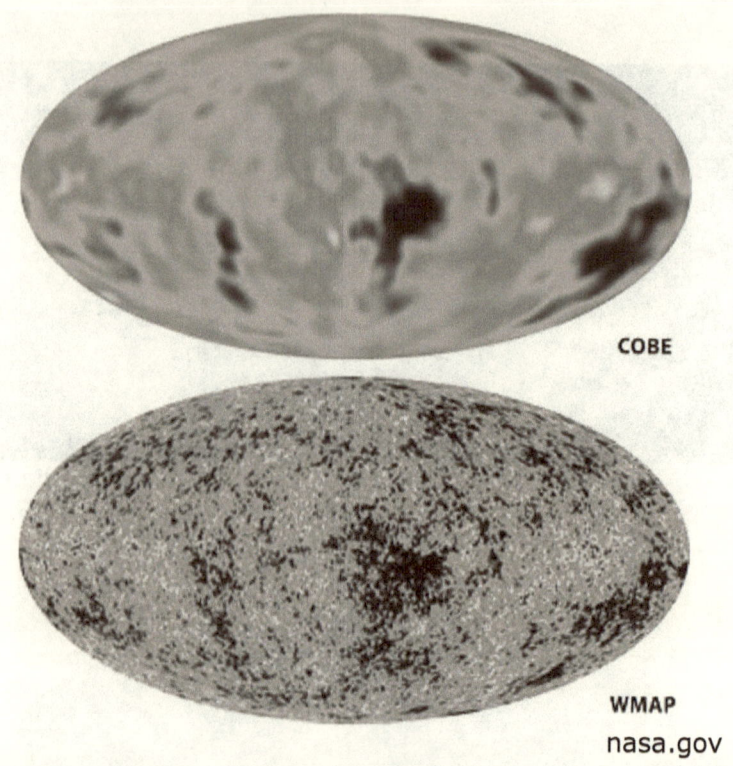

COBE

WMAP

nasa.gov

Figure II-11: Comparison between the anisotropy maps made by
COBE (1989) and WMAP (2001) satellites.

Smoot reported the fluctuations of intensity he observed in
units of temperature. Given the Big Bang Blackbody
temperature $T_{BB} \sim 2.7$ K, a deviation of the observed intensity
level can be converted to a deviation in temperature ΔT. The
fluctuations Smoot found were of the order $\Delta T/T_{BB} \sim 6 \times 10^{-6}$.
Thus the anisotropy was at the level of tens of microkelvins.

Later observations of the anisotropy by the Wilkinson
Microwave Anisotropy Probe (WMAP) Satellite in 2001 under
the leadership of Charles Bennett was said to refine and
corroborate the Smoot's discovery map (Figure II-11).

George Smoot's DMR findings, however, did not stop there.

64

He used his data to indirectly estimate the absolute sky brightnesses of the blackbody radiation at his three frequencies. A report on how this was done could not be located. However, his three data points fell right smack on Mather's perfect blackbody curve, providing an independent discovery of the blackbody (Figure II-12). Even if Mather's discovery did not stand, Smoot would still have a complete, stand-alone discovery: The blackbody *and* the anisotropy.

Figure II-12 (*overleaf*): In this composite diagram showing how many experimenters have independently verified the Big Bang blackbody, Smoot's three data points (x's pointed out by downward arrows) provide strong support for a well-defined peak of this diagram.

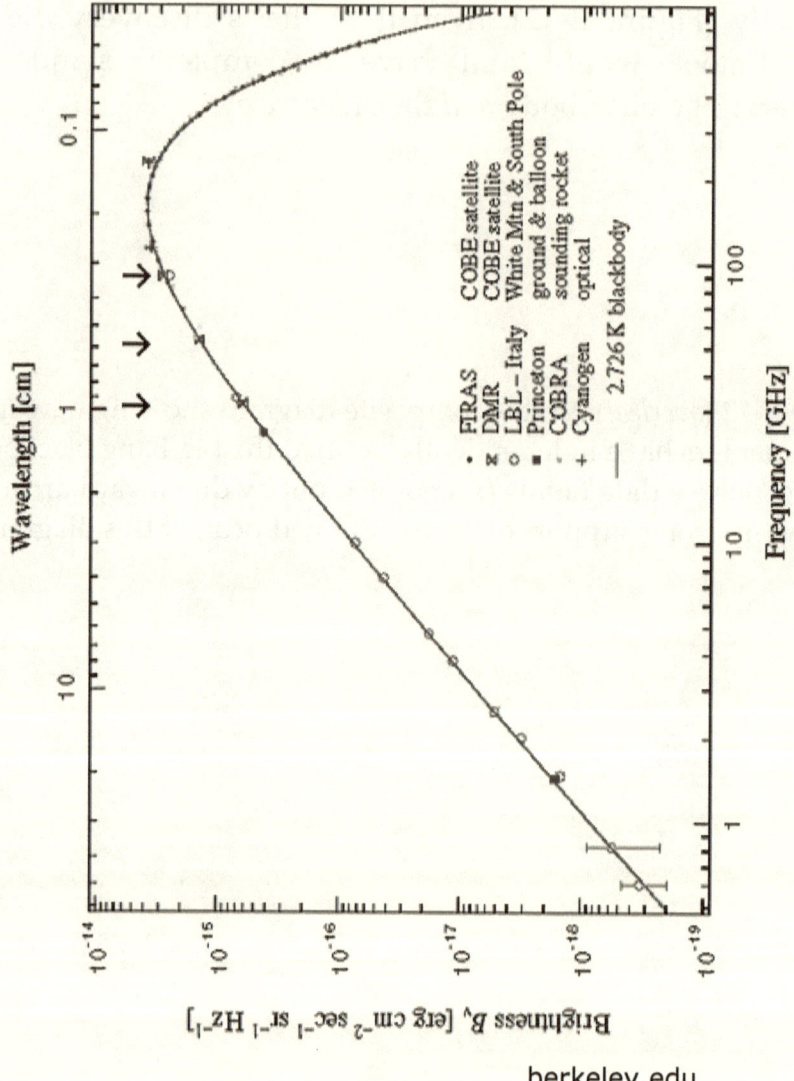

Figure II-12

CHAPTER II-4
Herbert Gush

Herbert Gush is not only not enshrined in the pantheon of Big Bang discoverers, he is today a little known figure in the arena's history. This should not be so. Perhaps the best way to describe his role today is to say he had, and has, the main discoverer John Mather's back.

ubc.edu

Herbert Gush

Herbert Gush at the time we are discussing was an illustrious University of British Columbia physicist and professor. He had joined the quest for the Big Bang blackbody early, operating on a very limited budget, mostly from the Canadian Space Agency. He had a small group of dedicated researchers to whom he was a good mentor. His approach was to launch an instrument package on a rocket, and make measurements during a portion of the flight of the rocket when it was clear of the Earth's atmosphere. Like other researchers in the area, he had his failures and partial successes, and was reaching the stage where success seemed to be close at hand.

Coincidentally, the progresses of Gush and Mather came to a head within months of each other. Gush was ahead and was on schedule to conduct his flight well ahead of the launch of the COBE satellite.

He lost out to NASA's COBE satellite team by barely a few months. Due to somebody's error, something went wrong during the vibration tests of his final instrument package, and that set Gush back by those months through no fault of his.

Herbert Gush's instrument – called COBRA – had some similarity to Mather's in that Gush too employed a blackbody calibrator (similar to Mather's external calibrator) and a single broadband horn-type antenna. But there were also important differences.

Refer to Figure II-13. The COBRA calibrator is a non-movable (unlike COBE) synthetic blackbody (Pyrex, as distinct from COBE's Eccosorb) shaped like a cone that permanently caps the mouth of a horn H2 identical to the sky horn H1. The sky radiation and the calibrator radiation are fed to the two input ports of a two-beam interferometer. The temperature of the calibrator is stepped until a null spectrum (null interferogram) is observed. When that happens, the temperature of the calibrator is the temperature of the blackbody in the sky.

The final flight of his instrument occurred on 20 January 1990. When the rocket achieved an altitude of 150 km, the measurements were begun. It achieved an apogee of 250 km. During the descent of the rocket, the measurements continued down to an altitude of 100 km. The measurement window was about 5 minutes.

Prior to launch, the instrument was tested in the laboratory to see that it was capable of obtaining a null spectrum. The sky horn (H1) aperture was capped with a simulated blackbody (Eccosorb) cone with controllable temperature. It was observed that a null interferogram resulted when the temperature of this blackbody in H1 matched that of the calibrator blackbody in H2.

THE COBRA INSTRUMENT

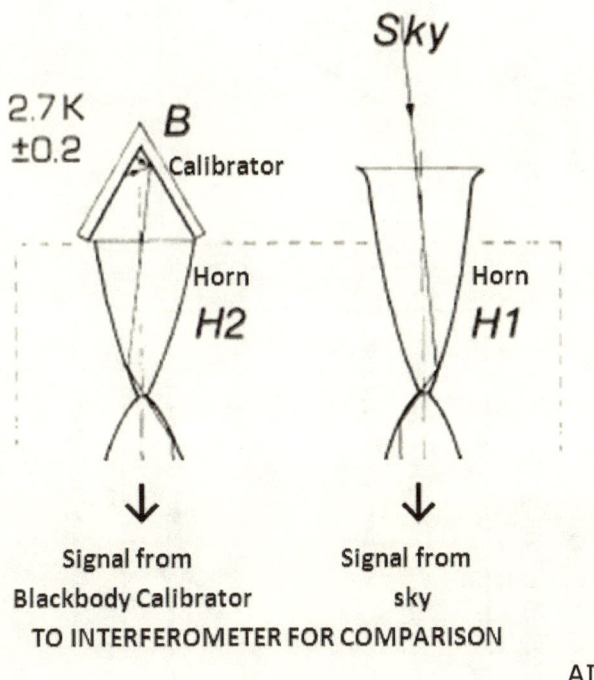

Figure II-13: The COBRA instrument measurement scheme.

Figure II-14 (*overleaf*): The Big Bang blackbody spectrum as discovered by Herbert Gush (COBRA rocket experiment) and John Mather (COBE satellite experiment), presented on approximately the same size scale for comparison.

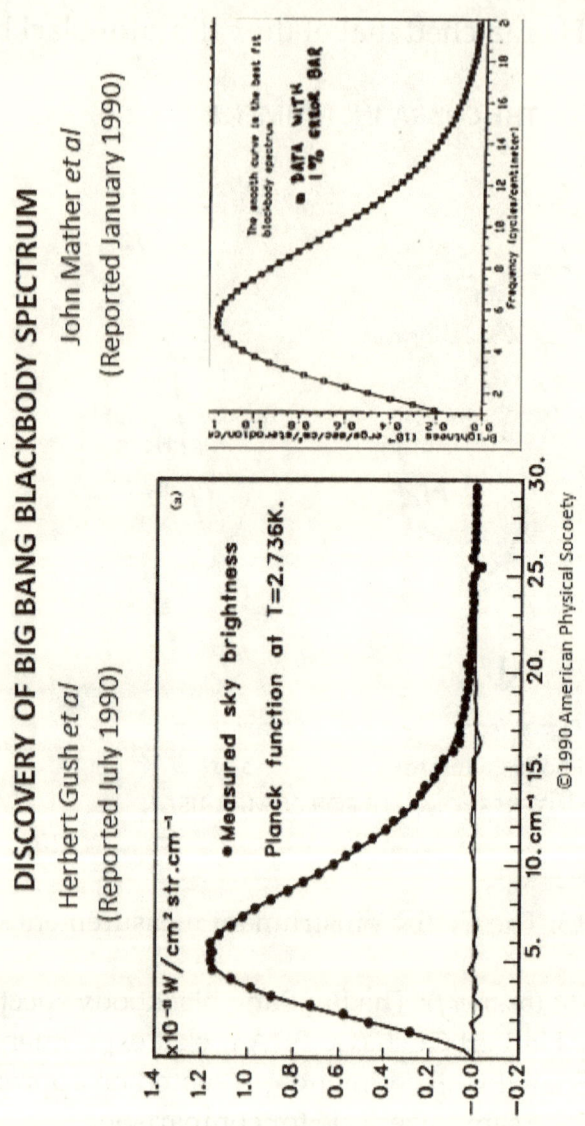

Figure II-14

Gush and his coworkers were able to report a nearly picture-perfect blackbody in the sky at 2.736 K. They claimed an even greater quality of measurement than the COBE team. Whereas Mather reported his result at the American Astronomical Society Meeting on 13 January 1990, Gush published his result in the 30 July 1990 issue of *Physical Review Letters*, considered the most prestigious publication vehicle in physics. The journal had received the Gush manuscript on 10 May 1990. Thus Gush missed out on the claim of prior discovery by a matter of a couple of months.

Herbert Gush gained neither fame nor fortune. But his discovery was important especially because it backed up John Mather's blackbody – perhaps open to disbelief for some hard-nosed experts because of its textbook perfect quality.

CHAPTER II-5
Perlmutter, Riess and Schmidt

Two Americans, Saul Perlmutter of the University of California at Berkeley and Adam Riess of Johns Hopkins University, and one American-born Australian, Brian Schmidt of the Australian National University, shocked the world by reporting in 1998 that the expansion of the Universe *is accelerating*. This was completely unexpected by the Big Bang cosmologists. Their lore held that the expansion of the Universe had to be slowing down due to gravitation of the material in it. The expansion had to be decelerating. The researchers were in fact trying to verify this very prediction.

wikipedia.org

Saul Perlmutter, Adam Riess, and Brian Schmidt

What exactly is this acceleration? What exactly is accelerating? It is space that is expanding at an accelerating rate. In Equation (I-5) we introduce the scale factor a. It is a linear measure of the expansion of the Universe with time. So we might write a as a function of time, $a(t)$. The expansion of the Universe can be symbolically expressed as

$da/dt > 0$ [*Friedmann- Lemaître -Hubble expansion*]

Then the acceleration of the expansion means, symbolically

$$d^2a/dt^2 > 0 \quad [\textit{Perlmutter-Riess-Schmidt acceleration of the expansion}]$$

So it is space that is expanding, and with it, the stars and galaxies. They are moving out farther and farther from one another. Acceleration of the expansion means that these objects are moving away from one another at a faster rate than the Hubble expansion told us.

The work leading to this phenomenal discovery was painstaking, lengthy and laborious. Even so, the result can be described in relatively simple terms.

In Chapter I-3 we introduced Hubble's law and the Hubble diagram. This diagram is universally seen as one showing an expansion of the Universe. Perlmutter *et al* were trying to extend this diagram to the high redshift region because of the theory that the high redshift objects represent more distant objects than those included in the preexisting Hubble diagram. These therefore are older objects of the Universe. If the expansion of the Universe is slowing down, these were the more likely objects to show this effect more definitively.

The researchers observed the light curves (time trace of the brightness) of a class of high redshift stars as they were going supernova (meaning, as they were exploding.) From these curves they calculated the absolute luminosities and the distances to these objects using the standard candle method. From this calculation it emerged that these supernovae are much dimmer than what an extrapolation of the Big Bang model of Hubble expansion to the high redshift supernovae would predict. This was interpreted as indicating that these newly observed objects were actually farther out (and hence dimmer) than where the Big Bang expansion scenario would place them.

From this it was concluded that the Big Bang Universe had to be expanding much faster than it was previously believed. The velocity of expansion was accelerating.

This result has been presented in many ways. One such diagram is shown in Figure II-15. Here the cluster of data points out to a redshift $z \sim 0.1$ represents the preexisting Hubble diagram. The data points to the right of $z \sim 0.1$ are the present discovery.

There are three theory curves in this diagram, calculated from the model of Big Bang expansion. The middle theory line represents the standard expansion scenario as befits the conventional Hubble diagram. The lower line represents a Universe that contains a great deal of matter that we did not know about (Dark Matter). This matter causes the Hubble expansion to slow down because of its gravitation, and thus decelerates the expansion. The upper curve represents a Universe that contains previously unknown energy (Dark Energy) which pushes on the matter and causes the Hubble expansion to accelerate. Clearly, the observed data points favor the upper line. Hence the discovery of the acceleration of the expansion of the Universe, and indirectly, Dark Energy.

How did this discovery strengthen Big Bang cosmology? It lent further credence to a quintessential Big Bang cosmology concept: Dark Energy. This is basically an intrinsic pressure of (or in) space of unknown origin that pushes out on everything and causes the Universe to expand faster than what the pull of universal gravitation would permit. In recent times attempts are being made to link the idea of *this* Dark Energy to Elementary Particle Physics and thence to String Theory. Such a linkage would lead to more entrenchment of Big Bang theory as a legitimate, hardcore scientific endeavor that also has a fascinating imaginarium aspect to it.

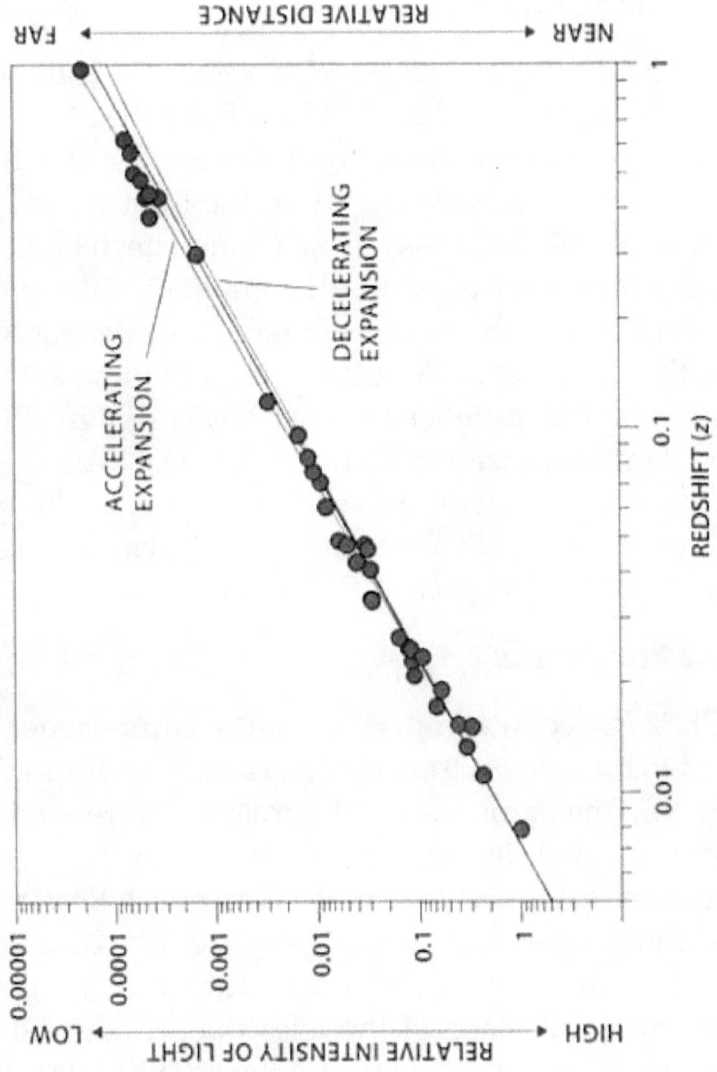

Figure II-15: Illustration of the discovery of the acceleration of expansion of the Universe.

CHAPTER II-6
BICEP2 Collaboration

II-6.1 Introduction

With the discoveries of the relic blackbody and the patchy structure of the sky predicted by Big Bang theory well under their belt, the natural order of business for the Big Bang cosmologists now was to set out on the hunt for the B-mode polarization swirls in the sky. This would clinch both inflation theory and the theory of primordial gravitational waves.

This task fell on the next generation, the one following that of Smooth, Mather et al. A cadre of bright young astronomers rose to the task. A number of experiments designed to detect the very faint polarization swirls in the sky were begun. Of these, the one that came to fruition first was the so-called BICEP2 experiment, BICEP being an acronym for *Background Imaging of Cosmic Extragalactic Polarization.*

II-6.2 Provenance

BICEP2 Project was a part of a large, multi-center program designed to hunt down B-mode polarization swirls in the sky. A number of the ground-based projects were centered on telescopes located in the South Pole. The BICEP program involved a refractor telescope in the Amundsen-Scott South Pole Station. The main work took place in the 2010 – 2012 timeframe.

Initially, dual polarized horn antennas were used to populate the focal plane of the telescope as the feed horns to make an image of the sky. This was the BICEP1 telescope. It did not succeed in positively detecting B-mode polarization swirls. It was then decided to use a different type of focal plane imaging instrumentation: slot antennas machined on a planar circuit board – hundreds of them. The antennas were alternately at 90 degrees to one another to receive two orthogonal polarizations (Figure II-16). Upon pointing the telescope to a region of the sky, this region would be imaged on the plane of the antennas. The antenna elements would then analyze the signals to generate a polarization map of that region.

The BICEP2 team leaders: (*From left to right*) Clement Pryke, James Bock, Chao-Lin Kuo, and John Kovac.

The initial design of this imaging plane by James Bock contained crossed slot antennas – two slot antennas intersecting in an X configuration. This was found to not work well. At this point Chao-Lin Kuo took over the function of designing the imaging plane.

It was determined that the above design had two problems. First, the isolation between the two crossed slots was poor. Second, the lead lines to the antennas on the said circuit board were interfering with one another.

After trying various solutions to these problems, Kuo settled on the design shown in Figure II-17(b). The antennas were now placed in an H configuration to eliminate the poor isolation problem. Some 500 such antennas were packed onto the imaging plane, compared to some 50 horn antennas of BICEP1. Each BICEP2 antenna was coupled to a highly sensitive bolometric detector that converted the electromagnetic energy to heat and measured the amount of heat. Also, the lead lines were strategically rerouted so as not to interfere with one another.

Chao-Lin Kuo then determined that in order to detect the B-mode polarization swirls in the sky, the instrument needed to have the ability to detect signals that are one part in 30 million compared to his main signal – the blackbody spectrum. By cooling parts of his instrumentation to cryogenic temperatures, he satisfied himself that this level of sensitivity was achievable.

Figure II-16 (*facing page*): An early diagram explaining the transition from the BICEP1 Telescope to the BICEP2 Telescope. Dual polarized horn antennas (two detectors per horn) were replaced by slot antennas (one detector per slot) on a printed circuit board. [TES = Transition-edge Sensor (detector element); SQUID = Superconducting Quantum Interference Device (detector)] (Image: Stanford.edu)

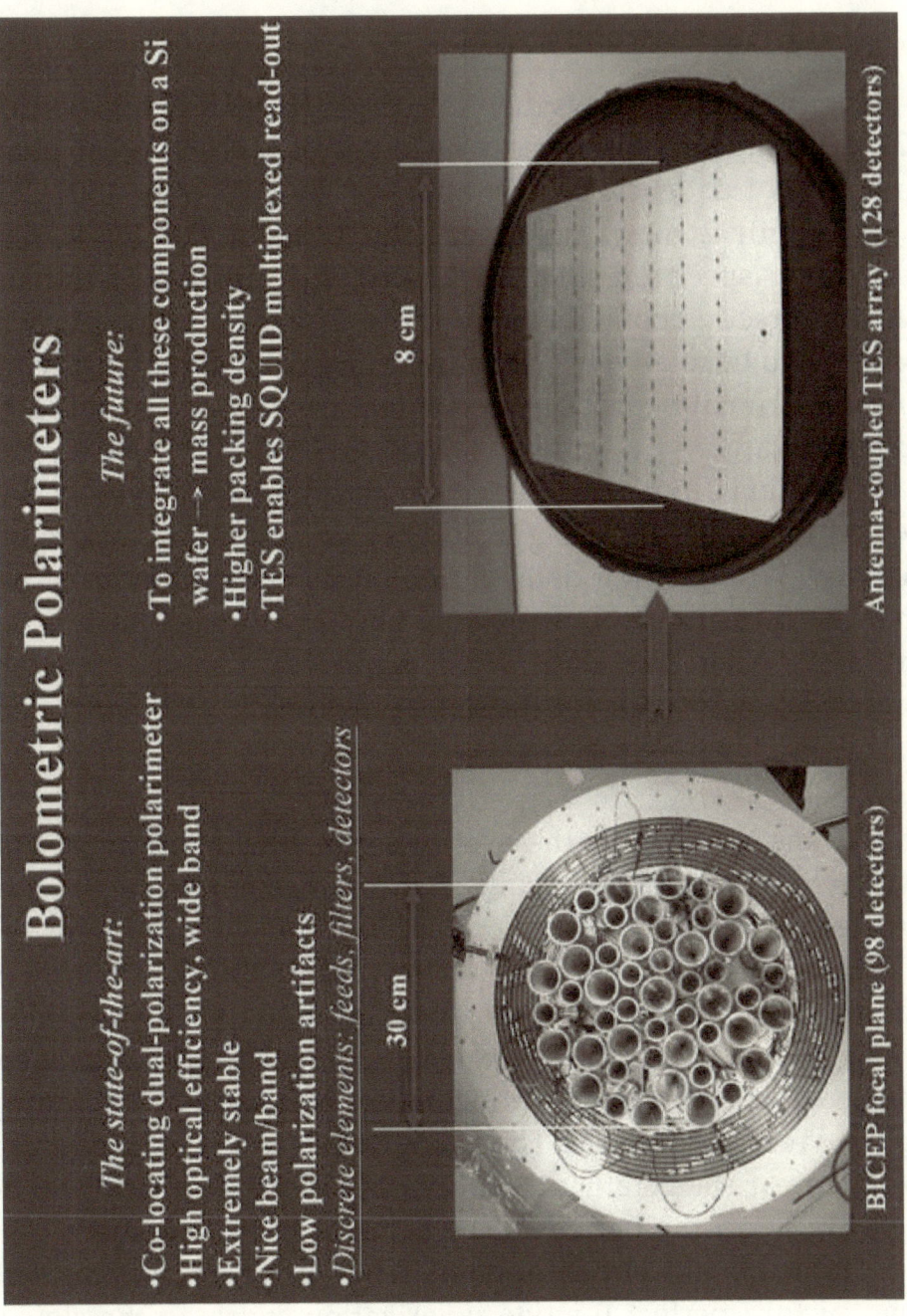

Bolometric Polarimeters

The state-of-the-art:
- Co-locating dual-polarization polarimeter
- High optical efficiency, wide band
- Extremely stable
- Nice beam/band
- Low polarization artifacts
- *Discrete elements: feeds, filters, detectors*

The future:
- To integrate all these components on a Si wafer → mass production
- Higher packing density
- TES enables SQUID multiplexed read-out

Antenna-coupled TES array (128 detectors)

8 cm

BICEP focal plane (98 detectors)

30 cm

Figure II-16

II-6.3 The telescope

Figure II-17(a) shows their telescope. It was a small aperture (26 cm diameter) telescope with a corresponding focal plane area. The frequency of operation of BICEP2 was 150 GHz (wavelength 2 mm), right near the peak of the 2.7 K relic blackbody spectrum. BICEP2 looked at a portion of the sky almost directly above the South Pole. This area of the sky was believed to be relatively "clean" – i.e., free of galactic dust. Thus any contaminating polarization signal from the dust would be minimized.

The data from this telescope were gathered over a long period of time – covering a number of South Pole observing seasons. While taking data, various tests were done to ensure that the measurements were true sky measurements, and not artifacts. After the data were acquired, various types of processing were done for the same purpose.

II-6.4 Success

At the end of all these efforts the BICEP2 team finally prepared to report the results. A Press Conference was set at Harvard University for 17 March 2014. It promised the disclosure of an important discovery. While the exact nature of the discovery was kept secret, people guessed that this would be about primordial gravitational waves. Great atmospherics were thus created and palpable anticipation was seeded, leading up to the Press Conference. The Conference was accompanied with great stagecraft, the BICEP2 leaders all wearing some type of uniform. A barrage of TV cameras pointed at them.

The team did not disappoint. They unveiled nothing less than picture perfect B-mode polarization swirls in the sky. A statistical certainty of 99.9997% of the discovery being correct was reported, reminiscent of the similar mind-boggling accuracy of 50 parts per million with the COBE satellite discovery of the relic blackbody spectrum. To appreciate the scientific enormity of this achievement, one should note that the polarization signals being measured in the sky here were probably a million times smaller than the signals reportedly measured by COBE.

The scientific establishment and the media went into a tizzy, trying to out-bloviate each other. It seemed that there were not enough superlatives in the English language to describe such an achievement by mere mortals. The evening newscasts around the globe capped this day of celebration.

However, the euphoria did not last long. Scientific groups from great centers of research such as Princeton University and Oxford University pointed out that these results could be explained in their entirety as polarization of CMB produced by galactic dust. Rather than being the deep background CMB polarization signal, the BICEP2 results may be showing the foreground dust polarization signal. The person that emerged as the leader of this dissidence movement is David Spergel of Princeton University.

So there was here no observation of gravitational waves and no confirmation of inflation. The rival teams made calculations to prove their point, and the Planck Collaboration contributed analysis of Planck satellite data to bolster this view. The BICEP2 team initially resisted this criticism vehemently, but eventually softened their stance.

After some back and forth debate, it was generally concluded by all parties concerned that the BICEP2 team had indeed discovered B-mode polarization in the sky, a great achievement in itself; but that no gravitational waves were discovered nor was inflation proven.

(a) BICEP2 Telescope

2.8 mm

(b) Configuration of slot antennas in the imaging plane

BICEP2 Collaboration

Figure II-17: The BICEP2 instrument.

The BICEP2 paper was eventually published in the journal *Physical Review Letter* in June 2014. The authors softened their stance that they had discovered gravitational waves, and allowed for the possibility that the entire B-mode signal was due to dust.

The leaders of the scientific establishment now moved in powerfully to blunt the blow on the BICEP2 team. They issued lofty homilies such as *This is how science is done* and *This is how science progresses*. The whole business was given a spin as science's Business as usual – and a very healthy business at that. The BICEP2 Team was saved harmless.

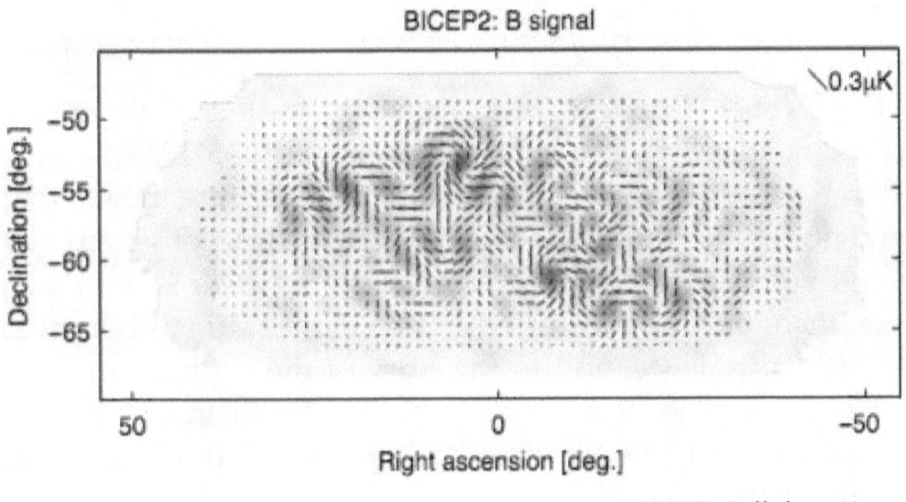

BICEP2 Collaboration

Figure II-18: Typical B-mode polarization swirls in the sky reported by BICEP2 Collaboration.

CHAPTER II-7
Garlands for the victors

And such are the magnificent chronicles of the great discoverers of Big Bang cosmology. When one reads their exploits, one cannot help but feel that here are no run-of-the-mill discoverers. These are a very special breed of explorers who are systematically deciphering the Mind of the Creator. And they are extending mankind's Book of Genesis in the process.

In 1978 Arno Penzias and Robert Wilson were awarded the Nobel Prize for physics for their discovery. They shared the Prize with Pyotr Kapitsa whose discovery was unrelated to that of Penzias and Wilson. Kapitsa received half the prize money. Penzias and Wilson received a quarter each. Their citation read

… for their discovery of cosmic microwave background radiation

The Nobel Committee for Physics stated that in the intervening period since their experiment (1965-1978), ample evidence had emerged from the work of other researchers that left no doubt that the two AT&T researchers had observed one spectral point in the Big Bang blackbody spectrum. With this, the names of the duo became enshrined in the history of physics, in the lore of science, and in the story of the civilization. And there is a tangible reminder of this in the historical plaque installed by the U. S. National Park Service at the location of their discovery.

John Mather and George Smoot would be likewise anointed in 2006, receiving an equal share of the prize money each. Their citation read

… for their discovery of the blackbody form and anisotropy of the cosmic microwave background radiation

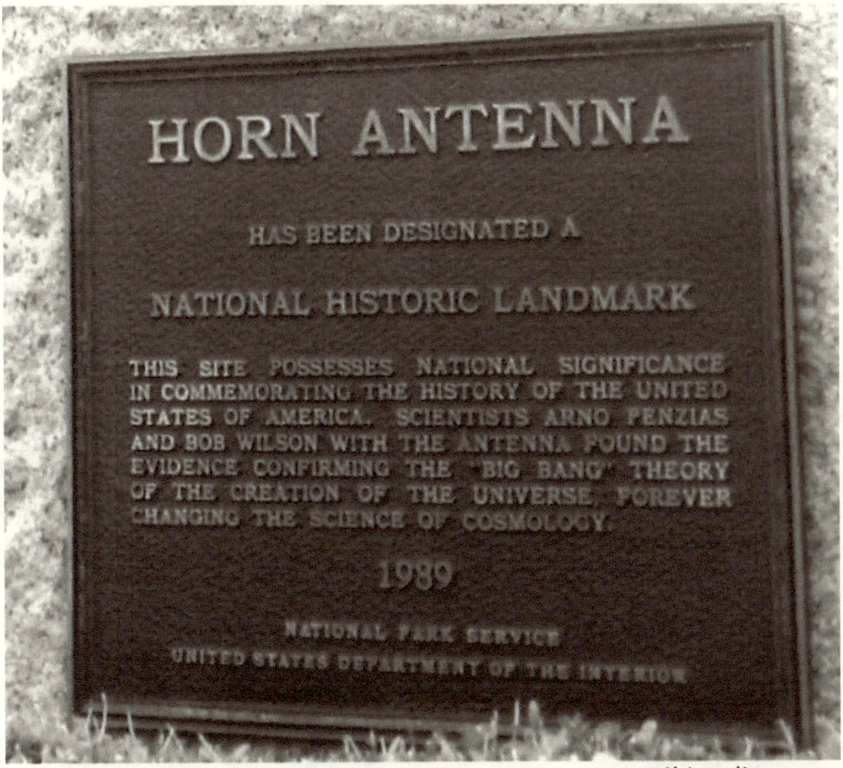

wikipedia.org

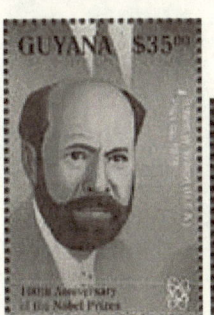

Plaque next to the Penzias-Wilson antenna in Holmdel, New Jersey, commemorating the historical discovery for posterity.

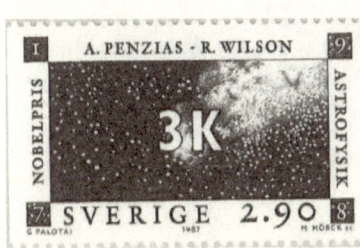

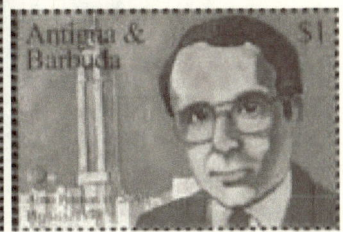

Commemorative postage stamps issued for Arno Penzias and Robert Wilson by Sweden, Guyana, and Antigua & Barbuda.

Saul Perlmutter, Adam Riess and Brian Schmidt received their Nobel Prizes in 2011, with the citation

... for the discovery of the accelerating expansion of the Universe through observations of distant supernovae

Perlmutter received half the prize money and Riess and Schmidt received a quarter each.

George Smoot and Saul Perlmutter were immortalized at their home institutions by the naming of streets after them – streets in the campus of the Lawrence Berkeley National Laboratory. John Mather was immortalized by the inclusion of a replica of the COBE satellite and one of his Nobel Prize Medal among the permanent exhibits of the Smithsonian Air and Space Museum.

lbl.gov

A street named after Saul Perlmutter in the campus of the Lawrence Berkeley National Laboratory. A cross street was named after George Smoot.

Needless to say, the Nobel Prize was neither the first nor the last in the string of awards that would flow towards these discoverers.

On the BICEP2 front, even as the dust vs. gravitational wave debate was raging, the Kavli Foundation moved forcefully to anoint inflation theorists Alan Guth, Andrei Linde and Alexei Starobinsky. They were awarded the 2014 Kavli Prize for Cosmology, which is considered something of a latter day Nobel Prize. After the awards were announced, the debate was resolved in favor of dust, thus pulling the plug from under inflation theory. However, in spite of the uncanny timing, the Kavli Foundation was careful not to cite the BICEP2 confirmation of inflation as the reason for the award. Instead they said that the award was being given because of the impact of the three men's ideas, as evinced by thousands of people working on these.

Barely a month after the announcement of the ill-fated BICEP2 discovery, team leader John Kovac of Harvard University was declared one of *100 Most Influential People* in the world by TIME Magazine. He was described as a pioneer. His citation was written by Brian Greene.

But around the edges of this brilliantly lit arena of golden laurels, there would be sighs of great sadness as well. Many expected Ralph Alpher, Charles Bennett, Robert Dicke and Herbert Gush to share in the Big Bang Nobel Prize bonanza. There were probably other hopefuls as well. In time they will all become footnotes of the history of Big Bang cosmology.

Images overleaf:
The names of American Nobel Laureates are literally carved in granite on the Nobel Monument in New York – another reminder that there is no going back.

nycgovparks.org

2003 A. A. ABRIKOSOV Physics A. J. LEGGETT Physics
P. AGRE Chemistry R. MACKINNON Chemistry
P. C. LAUTERBUR Medicine R. F. ENGLE III Econom
2004 D. J. GROSS Physics H. D. POLITZER Physics
F. A. WILCZEK Physics I. A. ROSE Chemistry
R. AXEL Medicine L. B. BUCK Medicine
E. C. PRESCOTT Economics
2005 R. J. GLAUBER Physics J. L. HALL Physics
R. H. GRUBBS Chemistry R. R. SCHROCK Chemistry
R. J. AUMANN Economics T. C. SCHELLING Economic
2006 J. C. MATHER Physics G. F. SMOOT III Physics
R. D. KORNBERG Chemistry A. Z. FIRE Medicine
C. C. MELLO Medicine E. S. PHELPS Economics

lbl.gov

88

THE BOOK OF APPLIED KNOWLEDGE

Even though Big Bang discoveries may seem to be the result of high-tech, complex, space age experimentations, at the core there is just old and well established science and technology.

Once one reviews these relevant fields, one becomes a savvy evaluator of the discoveries.

Book III is the key to everything here.

It gives the reader the confidence to dismiss out of hand the mountains of high-tech, high-math and high-science publications the Big Bang experimenters spinned.

They have wantonly destroyed so many of our great trees.

In theory there is no difference between theory and practice. In practice there is.

Yogi Berra

CHAPTER III-1
Antenna concepts

III-1.1 Introduction

In order to measure electromagnetic radiation from free space in the infrared and microwave frequencies, one will usually have to use suitable antennas to collect this radiation – that goes without saying. What needs to be said emphatically is that the antenna is the most crucial part of any such scientific experiments directed at quantifying this radiation. If this collection of radiation is not done right, what one does subsequently with the collected radiation is of no scientific relevance whatsoever.

For our discussion, we will re-familiarize ourselves with two types of antennas: parabolic reflectors and horns. The science of parabolic reflectors is what was used in the experiments that discovered the Big Bang blackbody spectrum. However, there may be an illusion that the science of horn antennas was used. So it is necessary to hold the basic distinction clear in mind.

III-1.2 Parabolic reflector antenna

The parabolic reflector (such as the ubiquitous satellite dish) is a metallic surface that follows a parabolic curve (Figure III-1(a)). The dish generally has a diameter that is far greater than the wavelength of radiation for which it is employed. Therefore one can discuss this in terms of a ray-tracing geometrical optics.

By geometrical property of a parabola, any ray that is incident on this surface parallel to the axis is brought to the focal point F upon reflection. So if a parallel beam of radiation (a plane wave) is incident on the dish parallel to the axis, the entire energy will be collected at F.

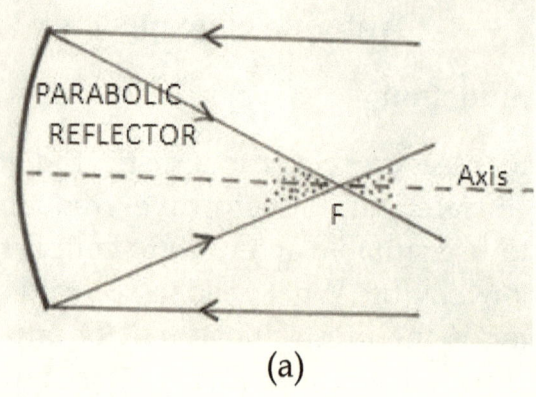

(a)

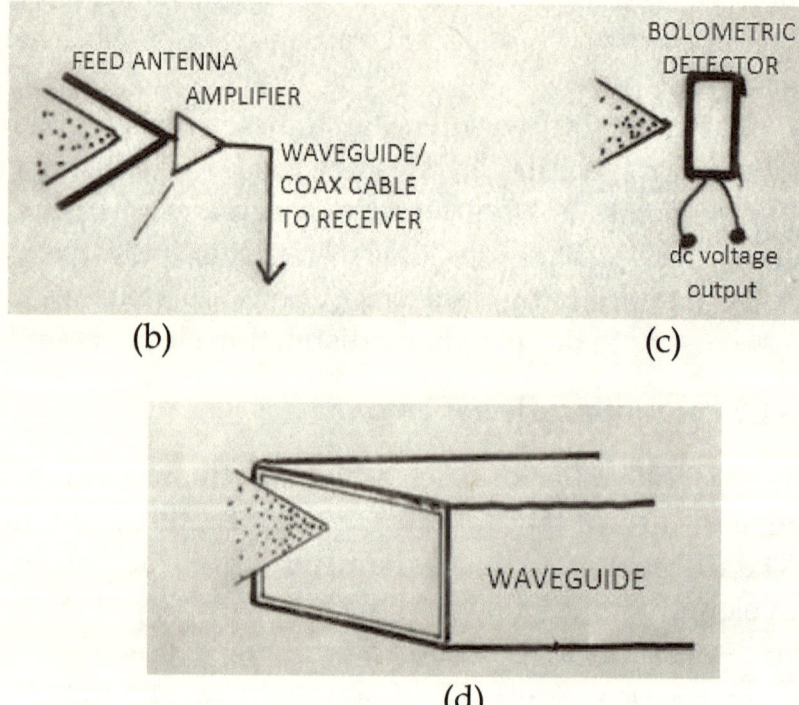

FEED ANTENNA

AMPLIFIER

WAVEGUIDE/
COAX CABLE
TO RECEIVER

BOLOMETRIC
DETECTOR

dc voltage
output

(b)   (c)

WAVEGUIDE

(d)

Figure III-1: (a) The parabolic reflector; (b) Feed for a parabolic
reflector; (c) Detection of received power at F;
(d) The waveguide feed.

For conceptual clarity, dotted shading is used to show the region of concentrated free space radiation that arises near F. If nothing stands in the path of this radiation, it will continue on back into space.

The geometric aperture of the parabolic reflector is the flat area that intercepts the incoming radiation. So it is roughly the flat circular plane enclosed by the rim of the reflector.

The parabolic reflector can be used effectively as long as the wavelength is small compared to the aperture diameter. Towards longer wavelengths (lower frequencies) its performance will degrade because scattering and diffraction, rather than geometrical optics, will control the collection of radiation. Towards shorter wavelengths (higher frequencies) the limitation of the parabolic reflector arises from how geometrically true it is as a parabola and how smooth its surface is. If the parabola is not true at the scale of the wavelength or if the surface is rough on this scale, the performance of the parabolic reflector will degrade.

Within the frequency range of applicability thus established for the reflector surface itself, the frequency coverage of the antenna as a whole will depend on how the radiation is collected at F and processed subsequently. We will return to this subject after discussing the horn antenna.

III-1.3 Horn antenna

The horn antenna, as the name implies, is a horn-shaped device (Figure III-2(a)) that collects radiation through its wide end and concentrates it as it travels down to the narrow end. The figure shows a conical horn (circular cross-section), but there may be horns with rectangular, square and other polygonal cross-sections as well.

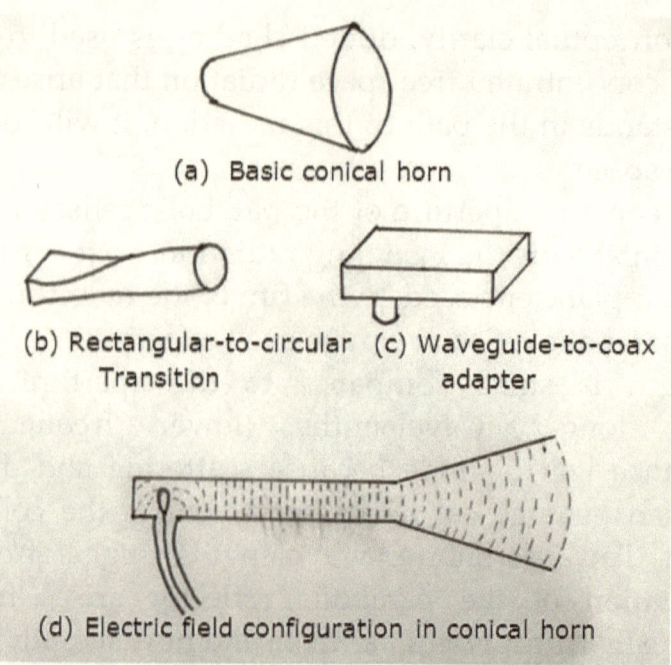

(a) Basic conical horn

(b) Rectangular-to-circular
Transition

(c) Waveguide-to-coax
adapter

(d) Electric field configuration in conical horn

Figure III-2: (a) A conical horn; (b) Transition from a circular pipe to a standard rectangular waveguide; (c) Waveguide-to-coaxial cable transition; (d) Overall electric field line (dashed lines) configuration of the electromagnetic wave in a conical horn.

A horn antenna, generally speaking, has an entry aperture (the flat area of the entry hole to the right) whose dimension is comparable to the wavelength (e.g., a few times the wavelength) of radiation it is to receive. The antenna operates over a frequency range that is fixed by this dimension – for the most part. So we can say that a horn antenna is bandwidth-limited.

If the application calls for a coverage over a much wider range of frequencies, then multiple horn antennas will be needed, at contiguous or overlapping frequency bands.

The frequency coverage of a horn antenna will also depend on the frequency coverage of all components behind it, up to the point of reception (detection).

III-1.4 Reception of radiation

This section will be a little descriptive and long-winded, but for a very crucial reason that will be clear later.

In the case of the parabolic reflector, free space radiation concentrated at the point F must be then received by a suitable antenna that is physically small enough so that it does not significantly block the path of the incoming radiation to the reflector.

This small antenna is called a Feed (Figure III-1(b)) – a terminology that arises from considering a transmitting antenna. In that case the small antenna sends or "feeds" radiation to the parabolic surface, and upon reflection there, a parallel beam of radiation is emitted to free space. While on this point, we can see that the feed should have a fairly wide beamwidth (see below) so that its radiation illuminates the full aperture of the reflector more or less uniformly. These concepts also apply to the parabolic reflector at reception.

So the feed we should place at F should be physically small and should have a wide beam. The bandwidth of the feed – its effective frequency range – will determine the bandwidth of the overall reflector antenna-feed system.

Just to give an idea: Depending on the frequency range of interest, the feed could be an electric dipole or a horn.

Once the concentrated radiation has been received by the feed, it is guided along a coaxial cable which is typically connected to an amplifier. After this amplification stage, the signal can be used for the intended purpose.

The purpose usually is communication, radar etc. The present application, using antennas to make high precision quantitative measurement of the intensity of radiation in free space, is a very specialized field. Here one could in principle consider two approaches: Detect (measure, quantify) the strength of the radiation past the feed or the amplifier; or have the concentrated radiation at F impinge directly on a bolometric detector placed right there (Figure III-1(c)). This detector will then generate a dc voltage proportional to the strength of the electromagnetic energy incident on the detector surface.

In the case of the horn antenna, the situation is different. We cannot here discuss a ray-optic type scenario. We must instead look at how the incident radiation travels along the horn. This is done usually by tracing the electric field of the wave as it enters the horn, following principles of electromagnetic theory (Maxwell's equations).

There are many types of horn antennas but we will consider only the conical horn which has a circular cross-section (Figure III-2(a)). The backport of the horn is a circle opening. But we must transition it to a rectangular port so that we can interface with a standard waveguide (which is a rectangular pipe of standardized dimensions.) The transition from the circular port to the rectangular port is done in a pipe that is designed from the principles of Microwave Theory – so as to smoothly convey most of the radiation leftward, with very little radiation reflected back towards the mouth of the horn (Figure III-2(b)). However, a sudden step change from a circular to rectangular pipe is possible, if designed appropriately. A stepped change from a circular to a square to a rectangular cross-section is also done.

So now the backport is rectangular and can match to a standard waveguide which can simply be attached with bolts at the flanges of the two components (Figure III-2(c)). Radiation collected by the horn will flow uninterrupted into the standard waveguide.

At the back of this waveguide is a properly designed waveguide-to-coaxial cable adapter (also standard.) The outer metal jacket of the coaxial cable connects to waveguide metal casing. The center conductor of the cable protrudes into the waveguide through a hole in the wall, and is attached to a teardrop-shaped metallic probe. The electric field lines (shown by broken lines (Figure III-2(d)) will now produce a bridge between the waveguide and the probe, and thus generate a signal (current) traveling down the coaxial cable. In this way the electromagnetic fields in the horn are converted to a current in a cable. One can now use this signal anyway one wishes.

If any one of the transitions discussed above is not designed properly, then a portion of the energy collected by the horn will be reflected back towards the aperture of the horn. This portion will be lost to the receiving electronics.

A form of the conical horn that we will encounter is called the corrugated horn. The corrugation refers to a series of circumferential grooves cut into the inside surface of the horn (Figure III-3). These grooves are cut following a prescription on the depths of the slots, the number of slots per inch etc. Clearly, the metal wall of the conical horn in this case has to be thick enough so that the grooves can be machined.

We will discuss presently the concept of antenna patterns. Referring to that concept, the corrugation of a conical horn results in three basic improvements:

(1) The E-plane and the H-plane beam patterns – which are different for the ordinary conical horn – tend to be equalized. Thus the cross-section of this beam is made more circular.

(2) The side lobes are lowered relative to the corresponding conical horn.

(3) The bandwidth of operation of the horn is increased relative to the corresponding conical horn.

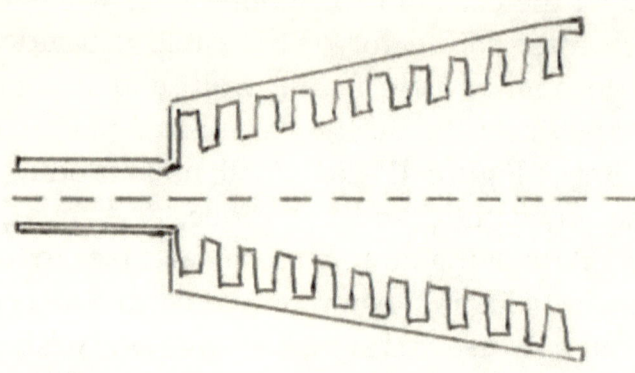

Figure III-3: The corrugated horn antenna.

III-1.5 Antenna pattern

The two main attributes that characterize an antenna besides its frequency bandwidth are its Gain and its radiation pattern. These two are related properties, but must be determined independently. Precision quantitative measurement of radiation requires the knowledge of both.

We will discuss the pattern first. This concept is at the very core of the measurement science in the satellite-based cosmology discoveries. One should bear in mind that this science of precision measurement with antennas is not widely practiced, and such requires special attention here.

The pattern of an antenna is essentially its directional sensitivity. The axis of a conventional antenna is the forward direction in which it is the most efficient receiver of radiation (and so this direction is also called the boresight of the antenna.) So if one has a distant point source of radiation and if one points the antenna directly at it, it will receive the maximum radiation. If one gradually turns (rotates) the antenna away from the source, the strength of radiation received will decrease first, and may then increase again – but always remaining below the boresight level. There are exceptions to this but we will not discuss them here.

If we make a trace of the strength of radiation power received as a function of the angle away from the source, we get an antenna pattern. The concept will become clearer if we discuss how the antenna pattern is measured.

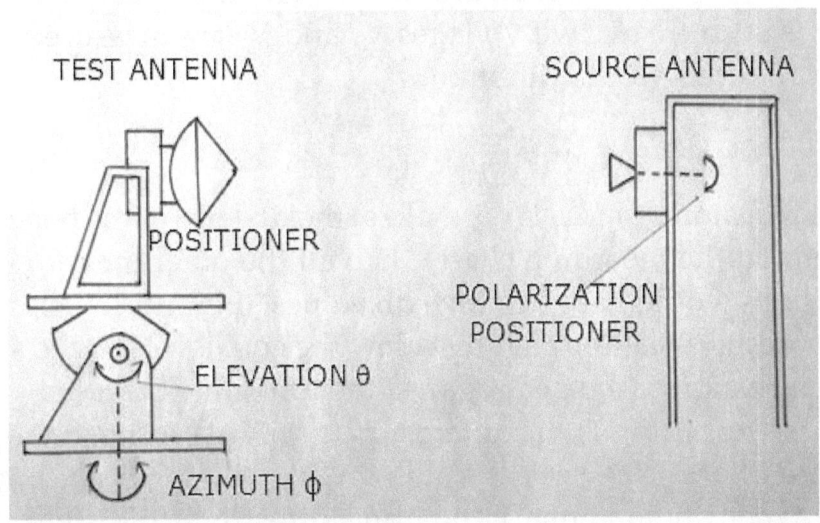

Figure III-4: The antenna range measurement set-up.

This is done in a setup called the antenna range (Figure III-4). A source of radiation, a transmitting antenna, is placed at a great distance (determined mainly by the wavelength of application) from the test antenna. This source must have a broad beam (we will come to the concept of beam presently.)

The test antenna is mounted on a pedestal so that it can be rotated in both the horizontal plane (azimuth coordinate ϕ) and the vertical plane (elevation coordinate θ.) The radiation received by the antenna is sent down a cable to a detector which converts the electromagnetic power it sees to a dc voltage. Before this happens, the frequency is often down-converted but this is not relevant for us. The antenna range equipment automatically rotates the antenna and records this voltage as a function of position angle θ or ϕ.

The dc voltage recorded represents the electromagnetic power received and may vary by many factors of 10 over the full angular sweep of the antenna. So it is customary to express it as decibels (dB) – a logarithmic scale based on powers of 10. The scale is such that if two voltages V_1 and V_2 are measured, then their dB levels would differ by

$$10 \log_{10} (V_1/V_2)$$

It is customary to set (normalize) the dB level at the boresight to zero on the recording chart. Then all the other measurement directions would have negative dB values. (Some technicians set this boresight reading slightly below the zero level so as to avoid a compression of the peak against the top of the chart.)

This recording is the antenna pattern. In a rectangular plot (Figure III-5(a)), its vertical scale is in dB, with the peak being at 0 dB when the antenna is pointing directly at the source. Its horizontal scale is in degrees of angle away from the source. The pattern is also sometimes presented in a polar diagram as shown in Figure III-5 (b).

Now we need to discuss this angle. The antenna pattern generally does not have cylindrical symmetry about the antenna axis. This means that the patterns obtained by rotating the antenna in the horizontal plane and the vertical plane are not the same. So we set up two orthogonal angular dependences: elevation θ which measures rotations in the plane of the paper, and azimuth ϕ in the perpendicular plane. The power received will thus be described as a function $f(\theta, \phi)$.

For essential reasons we will not go into here, it is necessary that the source illuminate the antenna aperture uniformly.

100

Another way of saying this is that the radiation from the source is incident on the antenna as a plane wave. This is one of the reasons the source has to be placed at a great distance, and has to have a broad beam. When not enough space is available for such a "long" range, there are other options we need not go into here.

Now let us look at the pattern as the antenna rotates in the plane of the paper. It is traced out customarily on a rectangular plot of the received power $P(\theta, 0)$ vs. the angle θ (Fig III-5(a), or a polar plot of the polar $p(\theta, 0)$ vs. the angle θ (Fig. III-5(b)). From this definition the meaning of $P(0, \varphi)$ and $p(0, \varphi)$ for movement along the φ axis should be clear .

Consider the movement along the θ axis, starting from the axis ($\theta = 0$). As you move away from the source, the power received decreases. It reaches a null, called the First Null. Then, it starts to increase and then decrease again. This up-and-down effect occurs because of constructive and destructive combining of the radiation that is distributed over the aperture of the antenna. As you rotate the antenna, the radiation over different parts of the aperture has traveled different distances from the source to F, and so – when combined at F – they interfere constructively or destructively. That is the essence of the undulating antenna pattern.

The pattern of the antenna between the first nulls is generally referred to as the Main Beam or the Main Lobe. It is characterized by a Beamwidth. There are different conventions of defining the beamwidth (expressed as an angle.) Sometimes the width between the half-power points (at approximately – 3 dB level from the peak) is taken as the beamwidth. This is the Half Power Beamwidth (HPBW). Sometimes the width between the first nulls is taken as the beamwidth. This is the Beamwidth between the First Nulls (BWFN).

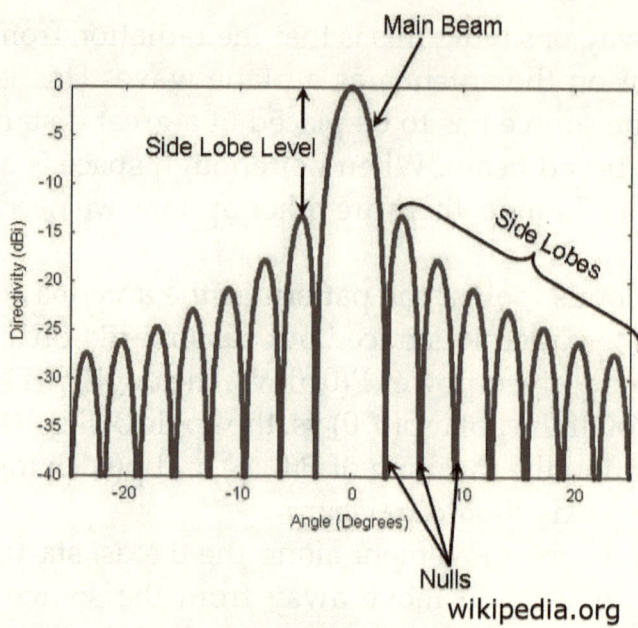

Figure III-5(a): Rectangular antenna pattern.

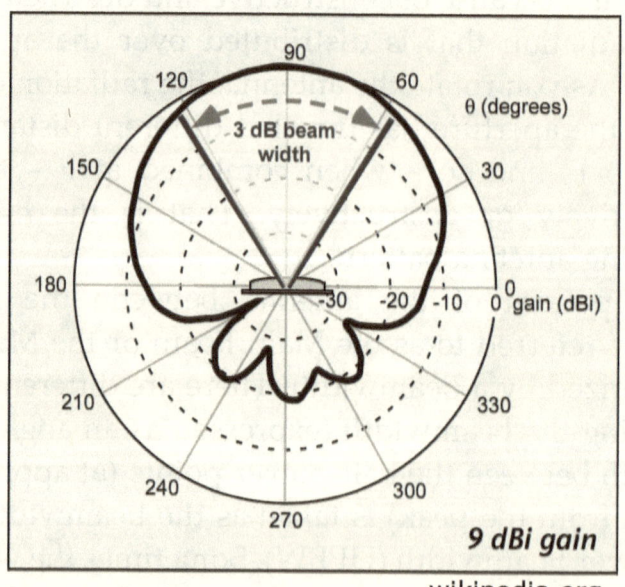

Figure III-5(b): Polar antenna pattern.

The other lobes in the pattern are referred to as the Side Lobes. The side lobe directly opposite the main lobe – the lobe in the back of the antenna – is called the Back Lobe.

III-1.6 Directivity, Gain and Effective Collecting Area

The antenna pattern tells us about the directional properties. The sharper the main beam (and so the lower the side lobes), the more directive is the antenna. Clearly, the pattern of an antenna that has no directivity at all will be a horizontal line on the rectangular plot and a circle in the polar plot. This idealized antenna, for the obvious reason, is called an omnidirectional or isotropic antenna. Other antennas are described in relation to this conceptual isotropic antenna. Thus, the technical term Directivity is defined by a relation:

$$D_v = 4\pi / \int_0^{2\pi} \int_0^{\pi} p_v(\theta, \varphi) \sin\theta \; d\theta \; d\varphi \qquad \text{(III-1)}$$

Here, the integral is that of the antenna pattern over the elemental solid angle $d\Omega = \sin\theta \; d\theta \; d\varphi$, to cover all space. The term 4π is the value of this integral for the isotropic antenna. The denominator is the value for the test antenna. So one can see how Directivity D_v will increase as the beam of the test antenna becomes more peaked and as the side lobes are lowered. One can also see that Directivity is essentially a "geometric" concept. The subscript v is included as a reminder that Directivity is a function of frequency. Directivity is a ratio and is commonly expressed in dBi (decibels with reference to isotropic.)

A concept very closely related to Directivity is Gain of an antenna. It is best understood by discussing an antenna in transmission, but the result also applies to antennas in reception.

Suppose we feed some electrical power to an isotropic antenna which then radiates equally in all directions. We measure the radiated power level at a great distance away from the antenna, and make a mark for this level on our chart recorder. Suppose then we replace the isotropic antenna with our test antenna, and rotate it 360 degrees. We record the maximum power level thus observed on the same chart. The difference (ratio) between the two power levels thus marked – expressed in dB – is the on-axis Gain of the test antenna.

Thus it is clear that Directivity and Gain can be different. The former is a geometric concept that does not include anything about the *efficiency* of radiation of the two antennas concerned; the latter is a practical concept that does. The efficiency concept arises because antennas can lose power through ohmic losses, losses due to surface roughness and other factors that are not represented in the power pattern alone.

When an antenna is in reception, a concept closely related to Gain is Collecting Area or Collecting Aperture A_v (θ, φ). It is the effective intercepting area an antenna presents towards radiation coming from the direction (θ, φ). If we multiply this area by the flux from that direction, we get an estimate of the total power received by the antenna from that direction. For a parabolic reflector, the Collecting Aperture in a slightly off-axis direction is close to the geometric aperture projected in that direction.

III-1.7 The concept of broadband antennas

The term broadband – as applied to antennas – is somewhat loose. In common usage, the bandwidth of an antenna is the range of frequencies over which it "works", according to some specific application. Thus, for example, for a communication antenna, at the lowest and the highest frequencies of its bandwidth, the performance of the antenna may have substantially degraded from its central design frequency. But as long as it can receive useful signal for communication, that frequency range is its bandwidth. Communication engineers might specify, as a hypothetical example, that – as you change the frequency - as long as the peak of the main beam remains above -10 dB level (with respect to the 0 dB level set for the central design frequency), then it is acceptable.

This usage cannot be carried over to scientific experiments where the antenna is being used for mensuration purposes and so where received signal strengths are the key. Here one must pay attention to the detailed antenna pattern at each frequency of interest. We are no longer satisfied with the fact that the antenna works, but we need to assure ourselves that its frequency response is fully quantifiable at each frequency.

Hence in the conventional usage of the term broadband, one is not so much concerned with the detailed antenna pattern as with the main beam region. The rest of the pattern is of interest only to the extent that this region not pick up spurious signals that interfere with the main communication.

In the scientific usage, the detailed quantitative pattern of the antenna at each frequency that will be reported must be known. Any analysis here must take into account the complete three-dimensional antenna pattern, or alternative methods that bypass this need must be developed and justified.

III-1.8 Antenna polarization pattern, axial ratio

We now introduce the concept of antenna polarization. Electromagnetic waves are made up of crossed electric field (E) and magnetic field (H). A linearly polarized antenna receives the maximum radiation energy when its direction of polarization is parallel to the electric field of the incoming wave, assumed to be linearly polarized. The antenna here is said to be co-polarized with the wave. If you rotate this antenna 90 degrees about its axis from this orientation, it should ideally receive zero radiation energy. The antenna is now cross-polarized with respect to the incoming wave.

If we refer back to the horn antenna of Figure III-2, we see that the electric field for that antenna is in the plane of the paper. This is the plane of polarization, and the horn is linearly polarized. The circular horn itself has no preferred polarization. It is the waveguide attached to its back that determines the polarization.

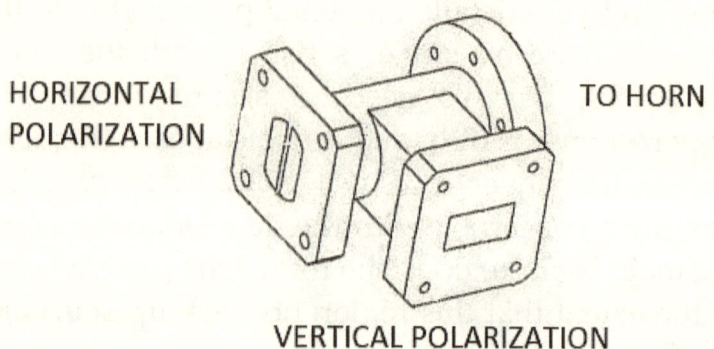

Image: wikipedia.org

Figure III- 6: Orthomode transducer

While on this point, we will also introduce the concept of a dual polarized horn for the purpose of later discussion. Here, instead of a single waveguide, one attaches to the circular backport of the horn an orthomode transducer (Figure II-6). This takes the radiation from the horn, however polarized, and splits its electric field into two orthogonal components (described in the figure as horizontal and vertical polarizations), emerging from two orthogonal waveguide ports.

In the case of the parabolic reflector antenna, the plane of polarization is determined by that of the feed horn.

If we measure the pattern of an antenna in a plane that contains the electric field, this pattern is called the E-plane pattern. If we measure the pattern in a plane containing the magnetic field (i.e., in the plane orthogonal to the E-plane), then we have the H-plane pattern. The two patterns are not necessarily the same. The three-dimensional antenna beam is not necessarily cylindrically symmetric about the axis.

Referring back to the antenna range (Figure III-4), the polarization properties of the test antenna can be studied by rotating the plane of polarization of the transmitting antenna which is linearly polarized in most conventional applications. Let us say that the polarization planes of the transmitting antenna and the test antenna are lined up with the former at zero polarization position angle. Here the two antennas are co-polarized. The test antenna is receiving the maximum power. Let us conveniently set this at -1 dB of our chart recorder (so that the peak of the graph will not be squished up against the 0 dB level.) The power level is plotted on a rectangular power-vs-position angle plot as the transmitter is rotated about its own axis. This generated the polarization pattern Figure III-7. A circular (polar) plot is also customary.

As expected, the power is maximum at the $0°$ position and minimum at the $± 90°$ positions, the cross-polarized positions. Ideally this power level should be zero, but it never is. The ratio between the maximum power and the minimum power is the axial ratio, a measure of how well polarized the test antenna is (assuming that the transmitting antenna is linearly polarized to a high degree.) In the figure the axial ratio is ~ 25 dB.

Any scientific instrument directed at measuring polarization must begin here, with studies of polarization pattern and axial ratio of the antenna element to be used. If that has not been done, any final results or discoveries reported should be discarded.

One final point: The incoming electromagnetic wave itself has its own state of polarization. It can be linearly polarized or randomly polarized (i.e., unpolarized) and anything in between. Instrument generated electromagnetic waves can also be circularly polarized.

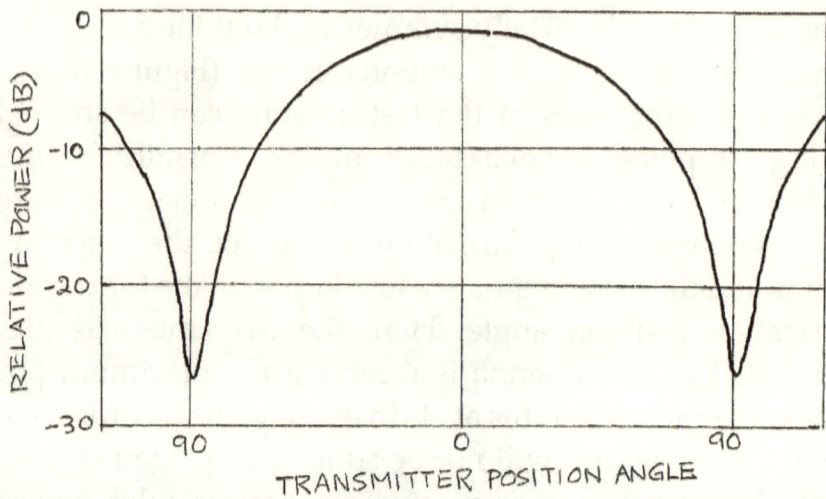

Figure III-7: Rectangular polarization pattern.

This is a simple discussion of polarization. In actual application, however, the subject becomes complicated by the introduction of geometrical considerations related to the actual orientation of the polarization vector in the sky.

III-1.9 Antenna near field, cross-talk and isolation

The near field of an antenna is the space around it where the electric fieldlines start on the antenna and end on the antenna. Beyond this region, these fieldlines decouple from the antenna and close upon themselves in free space. This is the freely propagating electromagnetic wave.

The near field of an antenna extends roughly a distance of one-third wavelength from it.

Two antennas with their boresights parallel should ideally not receive each other's transmission (except through the sidelobes.) But when they are in such close proximity that they encroach upon each other's near field, they no longer work as independent antennas. They interfere with each other's functioning. This phenomenon is quantified in terms of cross-talk or isolation.

Suppose that we transmit a power of x dBm (decibel above one milliwatt) through the first antenna. Then the second antenna in the above state will receive a fraction y dBm of this power. This is due to the direct coupling between the antennas. The isolation between them is (x-y) dB. The greater this number, the more independent the two antennas are.

Antennas in the near field of one another have their properties modified from when each operates ideally. These include virtually all properties of each antenna. One may choose to operate antennas in such configuration as long as one can tolerate the degradation of performance.

III-1.10 Antenna illumination

We will use a term for our purposes that is not exactly in accordance with the standard terminology in antenna studies: Antenna illumination.

Referring to the reflector antenna, this illumination could simply be a 2-dimensional map of the distribution of the electric field strength over the antenna aperture.

However, referring to the horn antenna, it would mean a 3-dimensional map of the same quantity within the volume enclosed by the horn.

Thus, if we later speak of an antenna in two different

situations having the same illumination (or not), this is what we are speaking of.

The pattern of an antenna has a direct relationship to the illumination of the antenna – whether we speak of reflectors or horns. The concept of the far field pattern – which is what the term pattern generally refers to - applies necessarily to points sufficiently far from the antenna so that the pattern does not change when one moves further away. Closer in, the pattern is called a near-field pattern and is dependent on the distance from the antenna.

Thus one cannot cap a horn antenna with an electromagnetically absorbing or emitting material or shroud a reflector antenna with such material and say that the cap/shroud covers the entire pattern of the antenna. This is a meaningless statement stemming from gross misunderstanding. We shall return to this subject.

CHAPTER III-2
Radiation field concepts

III-2.1 Intensity of radiation

Radiation field is the central concept in Big Bang experimental verifications. We will be dealing with such attributes of a radiation field as the energy, the frequency, and the state of polarization.

Energy in any free space electromagnetic radiation field is always moving at the velocity of light c. This movement can be characterized by the concepts of Intensity or Brightness, and Flux Density. Suppose we place a flat surface element in such a radiation field (Figure III-8). Then the Intensity I_v (θ, φ) of radiation at any point on this surface in the direction (θ, φ) from the surface normal is defined as the amount of energy incident on the surface per unit area of the surface per unit frequency band per unit time per unit solid angle subtended at that point around the direction (θ, φ).

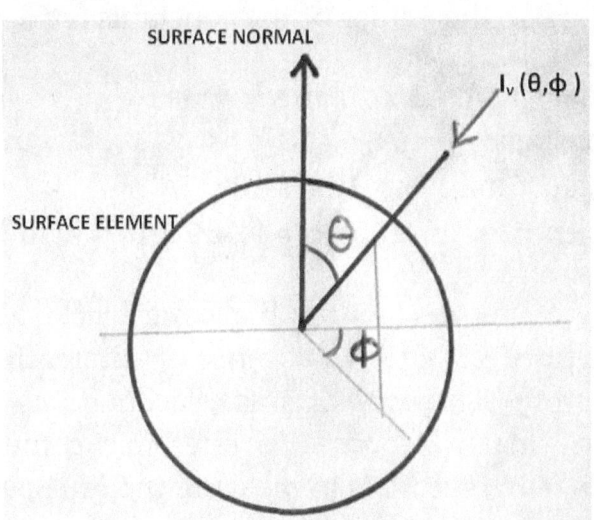

Figure III-8: Geometry for defining Intensity.

Thus in SI units the dimension of intensity is:

$I_v (\theta, \varphi)$: Joules sec^{-1} m^{-2} ster^{-1} Hz $^{-1}$

The quantity Joules sec^{-1} can be written as Watts. Other similar units can be used to describe intensity.

The Flux Density F_v of radiation is the total power crossing the surface element per unit area from all solid angles (i.e., from an entire hemisphere). Thus

$$F_v = \int_0^{2\pi} \int_0^{\pi/2} I_v (\theta, \varphi) \cos \theta \sin \theta \, d\theta \, d\varphi \qquad \text{(III-2)}$$

A particular type of radiation field of interest is the well-known blackbody radiation. This has the characteristic intensity spectrum $B_v (T)$ of radiation emitted by a 'blackbody' (which is literally black) at a temperature T while it fully absorbs any radiation incident on it at any frequency. This is a large subject. For our purposes here, we list some properties of this field in Table III-1. Here the values of the constants are as follows:

Velocity of light: $c = 3 \times 10^8$ m s^{-1}
Planck constant: $h = 6.63 \times 10^{-34}$ J s^{-1}
Boltzmann constant: $k = 1.38 \times 10^{-23}$ J/K
Stefan-Boltzmann constant: $\sigma = 5.669 \times 10^{-8}$ W m^{-2} K^{-4}

When we later discuss the Big Bang blackbody, we will be discussing pure radiation in absence of any matter. This means that the entire signature of this blackbody is its spectrum, without any material bodies to refer to. So the only way to measure its temperature is to measure the full spectrum. This is not a situation we encounter often, and needs thoughtful analysis.

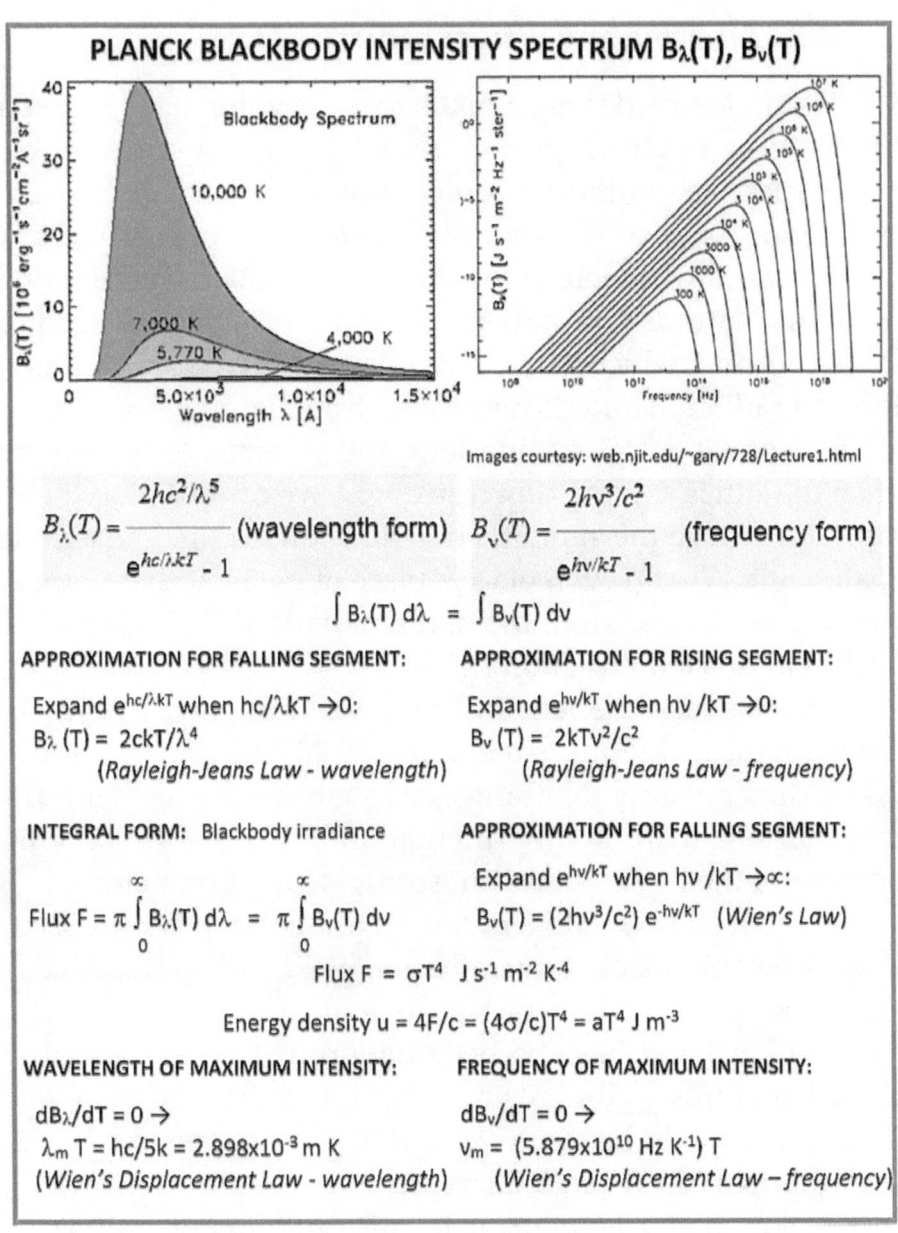

Table III-1: Some properties of the blackbody radiation field.

III-2.2 Directed and diffuse radiation fields

A radiation field is just what it says – a volume of open space filled with electromagnetic radiation in some range of frequencies. The radiation could have other properties such as continuous wave or pulsed radiation etc. The property that we are concerned with here is the directional flow properties of the radiation. Naturally, radiation can be crisscrossing a given point in space every which way. The two extreme cases of this are directed radiation and diffuse (or isotropic) radiation.

If we take a big reading lens and go outside in noontime when the sun is shining down strongly, we can position the lens perpendicular to the sun direction and concentrate the light to a small point. We can even place a piece of paper at that point and it may even ignite. Here the sun's radiation impinging on the Earth is a directed radiation field.

Next, we take the lens inside a well-designed greenhouse. The sunlight inside this greenhouse is all diffuse and, ideally, has no directionality. Nothing casts shadows in this light. If we hold our lens there in any orientation, there will be very little focusing of the light. This is an isotropic radiation field.

At a first level the relic radiation of Big Bang is proposed to be an isotropic blackbody radiation field. Integrated over all directions, the radiation has no overall directionality. This is how the Big Bang blackbody radiation in the sky is – an all-pervasive diffuse glow. The radiation is crisscrossing every which way – with the same strength in every direction. It is also randomly polarized (unpolarized.)

We note that whether our lens is outdoors or indoors, the intrinsic properties of the lens do not change. But what the lens does changes – precisely because its intrinsic properties do not change. It is the radiation field that changes.

Although radiative energy is always traveling at the speed of light, it is sometimes useful to define an energy density (a 'static' concept) of the radiation field, u (or its monochromatic value u_v.) This is just the amount of electromagnetic energy contained in a unit volume of space at any point at any instant of time.

For a directed radiation field (plane wave), a little consideration will tell us that

$$u_v = F_v/c, \tag{III-3}$$

whereas for an isotropic radiation field

$$u_v = 4F_v/c. \tag{III-4}$$

This distinction is thus relevant to the blackbody radiation field.

III-2.3 Antenna in different radiation fields

An antenna will measure different amounts of energy in a directed field and a diffuse field when the two fields have exactly same energy density at any given frequency. This is illustrated in Figure III-9.

Figure III-9(a) shows that if an antenna (shown by its radiation pattern) is lined up exactly to receive the maximum directed radiation, the energy is received only through the main beam. Figure III-9(b) shows that in a diffuse radiation field, energy will enter the antenna from every possible angle, the efficiency of reception being determined by the antenna's 3-dimensional pattern. Thus, even if the receptivity is low for wide angles off the boresight, radiation is entering over a large solid angle there. So the wide angle reception is not necessarily negligible in a diffuse radiation field.

In more precise language, the spectral power level per unit frequency bandwidth an antenna receives in any radiation field can written as an integral:

$$P_v = C \int_0^{2\pi} \int_0^{\pi} A_v(\theta, \varphi) \, I_v(\theta, \varphi) \sin\theta \, d\theta \, d\varphi, \qquad \text{(III-5)}$$

where $I_v(\theta, \varphi)$ is the intensity of the radiation and the integration is over the entire solid angle 4π. The subscript v indicates that the intensity is a function of frequency. The collecting area A_v we introduced in the previous chapter, and write it here more explicitly as a quantity dependent both on frequency and direction. The quantity C is a numerical factor that takes care of any mismatch between the antenna polarization and the polarization of the incoming wave.

Directed radiation field: In the case of an antenna lined up to receive maximum radiation from the direction (0, 0) from a directed field (Figure III-9(a)), $A_v(0, 0) = A_{v,max}$, the effective collective area in the boresight direction; and $I_v(\theta, \varphi) = 0$ except that $I_v(0, 0) = I_{vo}$. In this case

$$P_{v\_dir} = C \, I_{vo} \, A_{v,max} \qquad \text{(III-6)}$$

This shows that, ideally, we do not need to know the antenna pattern in detail. However, if we are concerned with stray radiation getting into the antenna from off-boresight directions, we do need to know the detailed pattern.

Isotropic radiation field: In the case of an antenna in an isotropic radiation field (Figure III-9(b)), $I_v(\theta, \varphi) = I_{vo}$ so that we can bring this factor outside the integral to obtain

$$P_{v\_iso} = C \, I_{vo} \int\int A_v(\theta, \varphi) \sin\theta \, d\theta \, d\varphi \qquad \text{(III-7)}$$

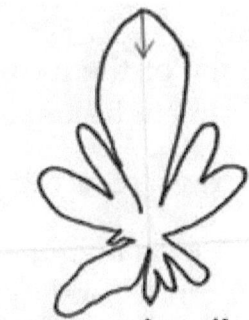

(a) The antenna pattern in a directed radiation field (arrow), lined up to receive the maximum radiation.

(b) The antenna pattern in an isotropic radiation field (arrows).

Figure III-9: An antenna in a directed and an isotropic radiation field.

This shows that in the isotropic case, we must have detailed knowledge of the directional properties of the antenna at each frequency. If we choose to circumvent acquiring this knowledge, it is up to us to justify that action.

The above necessity of having to know the detailed antenna pattern at each frequency is a very stringent, and nearly impractical one. But if one has made choices that bring one here, then such is how it is. It is not that *I* am imposing any needlessly

117

austere requirements.

The higher the accuracy of the measurement required, the more stringent this requirement becomes.

III-2.4 Antenna pattern vs. received power pattern

In space applications especially, one must be keenly aware of what an antenna pattern is and what a similar trace of received power is. As has been discussed, an antenna pattern is produced by an angular sweep of the antenna look direction past a distant point source that illuminates the aperture with a plane wave.

Thus, for example, sweeping an antenna likewise in an isotropic radiation field will produce no pattern. Sweeping it past a distant star should reproduce the antenna pattern itself. If the radiation field consists of both a distant star and an isotropic field, again, no pure antenna pattern will be reproduced.

III-2.5 Antenna Noise Temperature

The Antenna Temperature (variously called Antenna Noise Temperature and Noise Temperature) referred to the antenna terminal (before any amplifiers) is a thing engineers have defined for their convenience. It is not a real temperature of anything real. It is a measure of the irreducible "thermal" noise power at the antenna terminal that arises by virtue of the antenna being placed in a thermal bath – usually the heat from ground below and the sky above, and man-made heat all around. It is an equipment-specific number. The Antenna Temperature also depends on the ohmic losses in the antenna, but in our discussion we will assume that the antenna is lossless.

This noise power P_{noise} (resulting from all frequencies being received) that appears at the antenna terminal can be calibrated against a standard source of noise. Then we write $P_{noise} = kT_A$ to convert the power to the Antenna Temperature T_A (k = the Boltzmann constant). That is basically all there is to it.

Referring to Figure III-10, we will now examine the relationship of the Antenna Noise Temperature to the actual temperature field $T(\theta, \varphi)$ existing there. So, for an antenna pointing directly to the zenith, $T(\theta, \varphi)$ towards the front of the antenna would be the temperature of the sky. Towards the back of the antenna it would be the temperature of the ground.

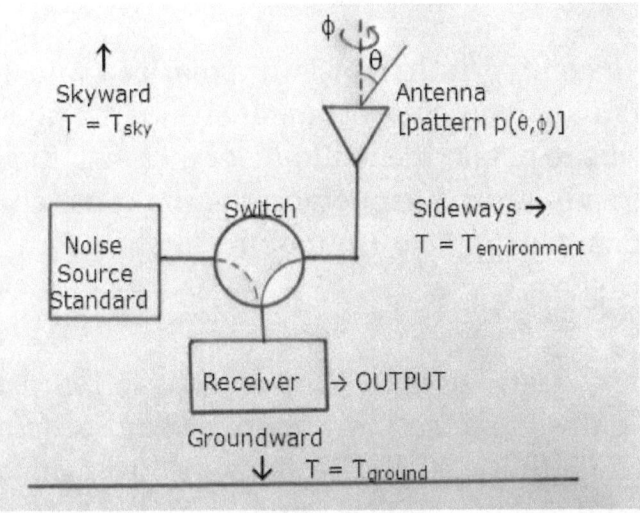

Figure III-10: Arrangement for measuring Antenna Noise Temperature.

By definition and by the looking at actual physical process, the Antenna Temperature is the average temperature seen by the antenna, modified by the polarization of the antenna:

$T_A =$

$$C \int_0^{2\pi} \int_0^{\pi} T(\theta, \varphi)\, p(\theta, \varphi)\, d\Omega \Big/ \int_0^{2\pi} \int_0^{\pi} p(\theta, \varphi)\, d\Omega \qquad \text{(III-8)}$$

119

where $d\Omega = \sin\theta\ d\theta\ d\varphi$.

Thus the polarization factor C is of crucial importance if we want to connect a measured T_A to the actual temperature T.

If the antenna is linearly polarized and the incoming radiation is unpolarized (or randomly polarized), then $C = \frac{1}{2}$.

If the antenna is linearly polarized and incoming radiation is linearly polarized and aligned with the antenna polarization, then $C = 1$.

If the above two linear polarizations are orthogonal, then $C = 0$.

If the incoming radiation is unpolarized and the antenna receives radiation in two orthogonal polarizations and adds these powers to obtain the total power incident, then $C = 1$.

If the environment temperature is the same all around the antenna, then we can take T out of the integral sign, leaving us

$$T_A = CT \tag{III-9}$$

Just to be clear, this is the Antenna Noise Temperature of an antenna totally immersed in a thermal bath of temperature T extending to infinity in all directions. If the antenna is linearly polarized and that radiation is unpolarized (as thermal radiation is), then $C = \frac{1}{2}$ and

$$T_A = T/2 \tag{III-10}$$

To measure the Antenna Noise Temperature, the noise power output from a standard noise source is adjusted to give the same receiver output as the noise power from the antenna. When this happens, the noise temperature of the source is the Noise Temperature of the antenna. It is clear that:

(1) The Antenna Temperature T_A and environment temperature T cannot be related without the knowledge of the

factor C.

(2)   The factor C is known only if the polarization state of the incoming radiation is known. To know the polarization state of unknown radiation, one needs a linearly polarized antenna whose plane of polarization can be continuously rotated at least 90 degrees.

(3)   An unknown radiation field cannot be determined to be a thermal radiation field without detailed and laborious measurement of the absolute intensity at multiple frequencies. To measure the isotropy of this radiation, one needs an antenna capable of pointing to a wide range of angles.

We note in passing that in the large majority of authoritative textbook expositions of Antenna Temperature, the factor C is omitted without any mention or discussion. This is because in almost all such applications the antenna and the radiation are linearly polarized and matched to each other. But this accepted and understood omission may easily lead an amateur user of the expression astray.

CHAPTER III-3
Derived antennas

III-3.1 Holmdel Horn

The antenna used in the Penzias & Wilson discovery (Figure II-1) is sometimes referred to as a horn antenna and has in fact the moniker Holmdel Horn. In actuality, it is very much a parabolic reflector antenna. It is actually a segment of a parabolic reflector as shown in Figure III-11.

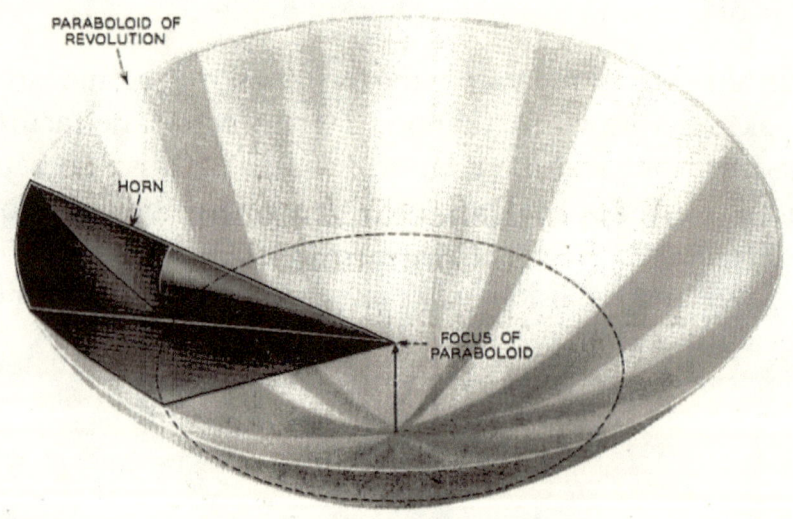

Figure III-11: The Penzias & Wilson antenna – although referred to as a horn – is actually a part of a parabolic reflector surface.

The radiation collected by this surface segment is directed to a feed horn located at the focus of the original parabola.

The feed horn is usually a linearly polarized antenna. This polarization is also the polarization of the overall antenna-feed system. If the feed horn can be rotated about its axis, then the plane of polarization of the antenna can be changed.

If the incoming radiation were unpolarized, this antenna

would receive only half the power. In order to receive the full incoming power, the radiation would have to be received in two orthogonal polarizations, and these powers would have to be combined in some way.

III-3.2 Winston Cone

The Winston Cone – so named after its inventor – is an antenna-like device developed for the main purpose of collecting and concentrating electromagnetic radiation from free space. Its object is to assess (or otherwise make use of) the *power level* of the radiation. It is not primarily concerned with any precision uses such as mensuration or imaging. It is sometimes called a heat trap or collector.

The Winston Cone – though referred to as a Sky Horn in COBE usage – is also takeoff on the parabolic reflector. In Figure III-12 we show a step-by-step process through which the properties of a parabolic reflector are used to construct a horn-like device. Sections of the parabola are put together to assemble this horn which will concentrate and deliver at the narrow end the power entering through the wide end. Our discussion is of course confined to a two-dimensional geometry. When one has assembled the actual cone, there will be rays that are not confined to two dimensions. This modifies the discussion somewhat, but generally speaking, the Winston cone does concentrate power in the manner shown in the figure.

Because of the manner in which the Winston Cone concentrates radiation, the emergent radiation at the back port is "scrambled". It cannot image an object it is looking at. This sacrifice of a property of the parabolic reflector results in an advantage: The Winston Cone can be used as a broadband collector of scrambled radiation.

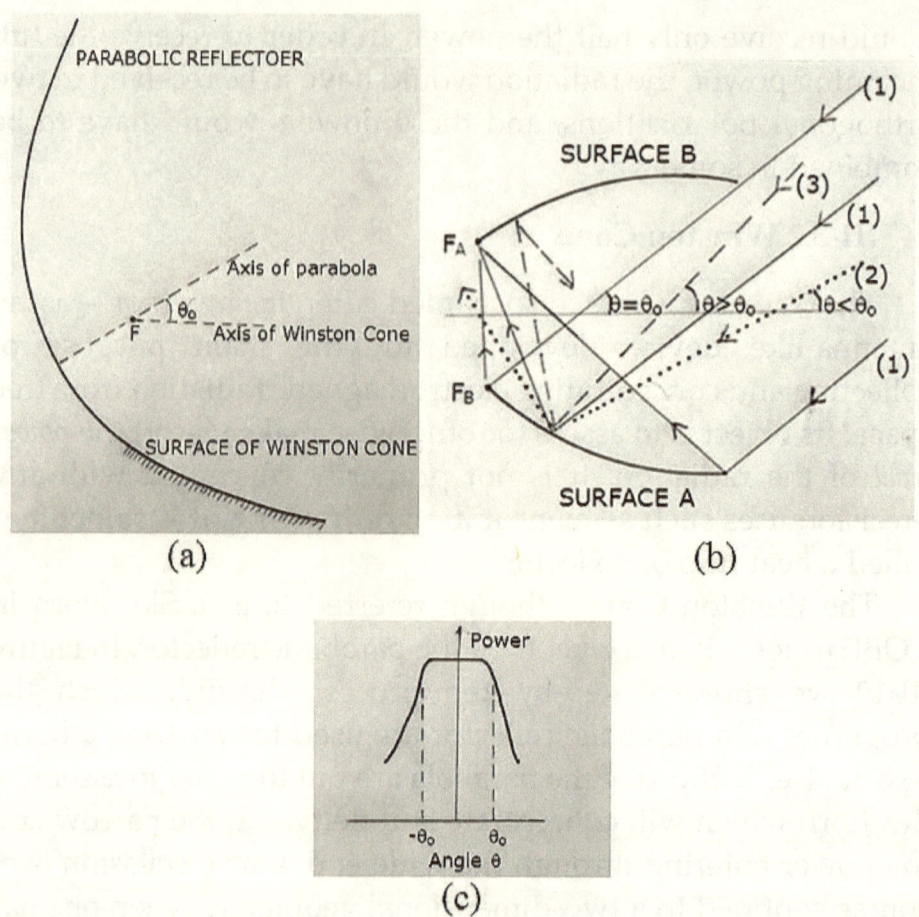

PARABOLIC REFLECTOER

Axis of parabola

F θ_0 Axis of Winston Cone

SURFACE OF WINSTON CONE

(a)

SURFACE B

F_A

F_B

$\theta = \theta_0$ $\theta \geq \theta_0$ $\theta < \theta_0$

(1)

(3)

(1)

(2)

(1)

SURFACE A

(b)

Power

$-\theta_0$ θ_0

Angle θ

(c)

Figure III-12: The Winston Cone (discussed here in 2-dimensions) is formed of the hashed segment of a parabolic reflector as shown in (a). All rays (1) pass through the exit port marginally. All rays (2) pass through the exit aperture. All rays (3) suffer two or more reflections and go back to space. Thus most of the radiation incident within an angle θ_0 of the axis of the Winston Cone pass through exit port. Beyond this angle, the reception falls off. Thus the Winston Cone has a flattop beam (c).

The three-dimensional Winston Cone is formed by rotating figure (b) about its axis. The two-dimensional geometry now no longer applies, but it still gives a good description of what happens.

124

As shown in the figure, the Winston Cone concentrates the radiation entering it over a solid angle around its axis, described by the angle θ_o. So, within this angle, the efficiency of reception is more or less the same. Outside this solid angle, the efficiency falls off rapidly. Thus is created a main beam that is flattop (or top hat) shaped. As long as the wavelength remains much smaller than the entry aperture, this flattop character of the beam as well as its shape and beamwidth should hold. This is how the Winston Cone is a broadband collecting device.

A low frequency cutoff of the Winston Cone bandwidth will arise when the wavelength starts to become comparable to the dimension of the aperture. A high frequency cutoff will arise when the roughness of the reflecting surface is of the same scale as the wavelength. The bandwidth of the Winston Cone is determined by these two limits, and can be quite broad.

So a flattop beam remaining unchanged over a broad range of frequencies is posited as the attractive property of the Winston Cone for the experiments directed at measuring the relic blackbody spectrum.

III-3.3 Winston Cone vs horn antenna

We now see that although the Winston Cone is based on the principles of the diagram in Figure III-1, the parabolic reflector, it looks remarkably like the diagram in Figure III-2, the conical horn.

Thus one may be tempted to attach to the back of the Winston Cone a circular-to-rectangular transition and then a standard waveguide, and in this way make tests and measurements. If one does that, one has made a foolish and fatal error.

What comes out of the backport of the Winston Cone is a concentrated beam of electromagnetic radiation traveling in free space. The wavelengths represented in this volume of radiation are much smaller than the size of the backport. This is not radiation that can be guided through a waveguide the size of the backport.

Referring again to Figure III-2, bottom, we can understand that if one were to attach a waveguide to the backport of the Winston Cone, some fraction of the radiation will attach to the probe and travel down the coaxial cable. Thus one will receive some power in this set-up. But this "leakage" power cannot be used to draw any conclusions whatsoever about the properties of the Winston Cone.

III-3.4 Satellite-mounted Winston Cone

Once we mount the Winston Cone on a satellite in some fashion, and then measure the beam pattern again (by rotating the satellite as a whole), the new beam pattern may or may not be the same as the pattern of the free-standing Winston Cone. Whatever the new pattern is, it is now the operative pattern. The pattern of the cone alone is of no consequence anymore. And it is the new pattern that must have all the properties the experiment calls for.

It follows that the Winston Come must be mounted on the satellite in a way that its beam properties remain unchanged from when it stood alone. Or the beam properties should be measured on the ground with the Winston Cone in the satellite-mounted configuration.

CHAPTER III-4
The science of calibration

III-4.1 Purpose of calibration

Calibration in the present context refers to providing a way for the observing system to compare the measurement of power from an unknown source to the power level from a source of known strength. By comparing the two power levels the former power level can be quantified as precisely as the accuracy with which the latter power level is known.

It is, however, most important that one here compares apples with apples and oranges with oranges – electromagnetically speaking. This is the special aspect of calibration we are interest in here.

There are basically two types of calibration signals: free space radiation from a known source and guided radiation (as in a waveguide or a coaxial cable) from an electronic device. Let us illustrate this with reference to a parabolic reflector antenna.

III-4.2 Open source calibration

In Figure III-13 an antenna is steered to alternately look at an unknown radio source X and a known source S of radio signal. The source S could be another astronomical source whose radio brightness has somehow been established well. It could also be a satellite which radiated power is known accurately. Let us say the source S is known to generate at the receiver output a power level which we know from the properties of this source to be -55 dBm. The source X is observed to have a power level 15 dB below source A. So the power level of the unknown source is -70 dBm.

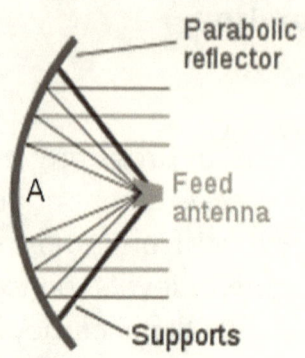

Parabolic reflector

Feed antenna

A

Supports

S * calibration source

X * unknown source

Figure III-13: Open source calibration: An antenna is sequentially pointed to the unknown source X and the known source A.

This is the simple way to look at calibration. Since X and S illuminate the antenna aperture in exactly the same way (in this case, with a plane wave parallel to the parabola axis), resulting in the same electric field geometry at the feed horn aperture, we are comparing apples with apples.

Sometimes it is convenient to have a known source of a "test signal" mounted at the apex A of the antenna. A small horn radiates energy towards the feed horn. This method does not reproduce the radiation field incident on the reflector, but reproduces the signal from the feed horn to the receiver.

III-4.3 Electronic source calibration

The signal from the feed is usually amplified right at the backport of the feed with an amplifier mounted there. Past this amplifier, the amplified signal flows down a cable to the receiver, usually located either right at the focus or in the observatory house.

An electronic reference signal of precisely known strength

can be injected into the path by means of a signal coupler. It is injected either before or after the amplifier. Thus it tests the signal path from the feed to the receiver. The receiver system is alternately switched to the source and the reference signal. Now we can determine the source strength in the same way as above.

Since both the source signal and the reference signal are traveling down a waveguide or a coaxial cable in exactly the same way with the same electric field geometry, we are comparing oranges with oranges.

II-4.3 Dummy source calibration

This is not at all a common calibration technique, but it is crucially important for our discussion. For proper context we will discuss this case referred to a "horn" antenna of the Winston Cone type, looking at an unknown sky signal.

As shown in Figure III-14, the calibration signal here comes from a reference blackbody which caps the mouth of this horn in some fashion. The idea is that this cap, depending on its temperature, emits a precisely known radiation spectrum (a blackbody spectrum) into the horn. It is assumed in these studies that the illumination geometry (optical ray geometry) inside the horn is the same whether the horn looks at the sky or either of the caps. That being the case, we are presumably comparing apples with apples. Since the power from the cap is precisely known from its temperature, the unknown sky signal can be quantified by referring to the cap signal.

This assumption is wrong for several reasons which are best discussed when we come to the actual applications. This is not a valid calibration technique at all.

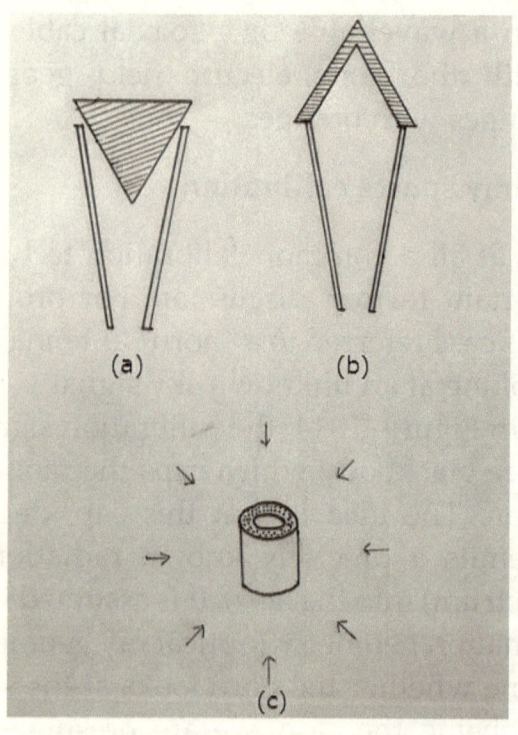

Figure III-14: Dummy-source calibration: (a) and (b): A cap made of a perfect blackbody material covering the aperture of the horn serves as a reference source. (c) When this cap is removed and the satellite-mounted horn looks at free space, it receives radiation from all directions, depending on its radiation pattern.

CHAPTER III-5
Temperature concepts

III-5.1 Temperature is the key

It is necessary for the reader to appreciate that the concept of temperature is the key, directly or indirectly, to all Big Bang experimental evidence. This concerns not only the cosmic blackbody, but also the age and the history of the Universe, the CMB angular power spectrum and so on. In the process of developing this body of evidence, the very physics definition of temperature was corrupted, and replaced by a Big Bang-ordained "temperature". This false temperature was then treated as temperature in the normal physics sense. This *ad hoc* "temperaturization" of observed intensity is a Big bang stratagem that we must remain constantly aware of.

III-5.2 Definition: Temperature of radiation field

A radiation field in space does not have a temperature assignable to it in any sense of physics. The concepts such as Antenna Temperature (referring to an antenna immersed in the field) are definitions of practical convenience. The only instance where we can speak of a temperature of the radiation field of infinite spatial extent in empty space is when this field is a blackbody radiation field with intensity B_v. This temperature T is defined in Table III-1.

If one makes a single frequency measurement in a known blackbody radiation field and finds that the observed intensity I_v at that frequency v is slightly different from the B_v expected in the blackbody spectrum, one can assign a temperature differential ΔT to the intensity differential $\Delta B_v = I_v - B_v$ as follows:

$$\Delta T \approx \Delta B_v / (\partial B_v / \partial T)_v,$$

where the term in brackets is calculated from the expression for B_v (Table III-1). This is the one and the only way we can speak of spatial temperature fluctuations in a radiation field.

The result in this case must consist of the maps in the sky for both T and ΔT, showing any correlation (or lack thereof) between the two fields. Furthermore, if ΔT is determined for multiple frequencies, all substantially within the blackbody spectrum, then ΔT must be the same for all frequencies for the same location in the sky. This would be a crucial test.

If the radiation field is determined to be a non-blackbody, no temperature of any kind can be assigned to it. If one assigns a temperature to a non-blackbody radiation field as a handy way of discussion, we must be on guard that it not be used later as a true temperature in the physics sense.

CHAPTER III-6
Satellite electromagnetic measurement theorems

We will now summarize the preceding discussions in the form of a set of *ad hoc* theorems that have specific application to the satellite electromagnetic radiation measurement experiments in cosmology. These theorems have the full force of basic scientific principles.

The following selected formulations – which will be patently obvious to antenna-and-microwave practitioners – address the main misunderstandings of relevant science by the Big Bang satellite experimenters.

THEOREM A: <u>Satellite-mounted antennas</u>

The on-axis gain and the three-dimensional pattern for a free-standing antenna (as in an antenna range) and a satellite-mounted antenna may differ significantly.

Corollary A-1: Absolute quantitative measurement of the intensity of isotropic radiation in the sky cannot be made without knowing the on-axis gain and the detailed three-dimensional pattern of the satellite-mounted antenna.

Corollary A-2: Communication satellites and meteorological satellites are very different from scientific radiation measurement satellites in this respect: In the former case the overall pattern (e.g., sidelobes and backlobe) is not so important.

THEOREM B: <u>Antenna in different radiation fields</u>

The same antenna placed in a directed (plane wave) radiation field and a diffuse (isotropic) radiation field – both fields having the same spectral energy density – will collect substantially different amount of net energy at the antenna output port.

Corollary B-1: The two spectra at the output port will have different shapes, dependent on the overall pattern of the satellite-mounted antenna.

THEOREM C: Pattern measurement concepts

For proper characterization of the antenna pattern, the aperture of the test antenna when pointed at the source must be uniformly illuminated by a plane wave.

Corollary C-1: One cannot measure the pattern by illuminating the aperture with a beam whose diameter is the size of the aperture or smaller.

Corollary C-2: One cannot expect to reproduce the ground-measured antenna pattern by sweeping the satellite-borne antenna past a point source (the moon, e.g.) if there is also present an isotropic radiation field of comparable strength.

Corollary C-3: If the ground-measured antenna patterns *are* reproduced faithfully in such an experiment, then there is no such isotropic radiation field present in the sky.

THEOREM D: Guided flow of radiation

For proper characterization of a scientific experiment, the guided flow of radiation in structures such as a narrowing horn and a horn-to-waveguide transition must follow prescribed rules of optics or microwave theory – as the case may be.

Corollary D-1: One cannot mechanically slap the exit port of antenna on to the flange of a waveguide and proceed to acquire scientific data. The two components must be electromagnetically compatible.

Corollary D-2: If two otherwise-identical horn antennas of circular cross-section end in the exit port with a circular cross-section and a rectangular cross-section, respectively, then these two antennas are completely different with respect to gain and pattern; the degree of polarization, the spectral energy density, and the cross-sectional shape of the beam that emerges from the exit port. These two antennas are not interchangeable.

Corollary D-3: A handy-dandy metal funnel cannot be used as a radiation collector in scientific measurement experiments. The collector must be an antenna in the scientific sense.

THEOREM E: <u>Comparison measurements</u>

Comparing measurement of sky radiation with some type of a satellite-borne calibrator radiation cannot obviate the need for the knowledge of the satellite-mounted antenna pattern.

CorollaryE-1: Therefore an interferometric subtraction of the energy at one output port (of an antenna looking at the unknown sky, e.g.) from that at another (a second identical antenna corked by a reference blackbody, e.g.) cannot result in a null spectrum even if the sky were a blackbody at the same temperature as the reference blackbody.

THEOREM F: <u>Circumventing detailed knowledge of antenna patterns</u>

Over the entire frequency range of a spectrum desired to be measured, the known on-axis Gain and the known antenna pattern must remain exactly the same.

Corollary F-1: Any departures from the above will result in unknown instrumental distortion of the unknown spectrum being measured.

135

Corollary F-2: Variation of the sidelobes below a certain specified level may be tolerated in this case, but only after an error analysis for the same has been made.

THEOREM G: <u>Differential sky measurements</u>

When differential measurement between two portions of the sky is made using a narrow antenna beam (to construct an intensity skymap, for example), it is not necessary to know the overall radiation pattern.

Corollary G-1: Such differential maps should not then be manipulated to indirectly obtain absolute intensities in the sky, and then to claim high precision for these intensities.

THEOREM H: <u>The main beam</u>

The main beam of an antenna is its main directional attribute. A correctly designed antenna always reproduces the major portion of the main beam faithfully. Any further development efforts are needed only to improve the depth of the first nulls and the levels of the sidelobes and the backlobe, and VSWR characteristics.

Corollary H-1: If the peak portion of the main beam turns out to be distorted, the antenna has not been correctly designed in the first place.

Corollary H-2: If the peak portion of the main beam of a satellite-mounted antenna devoted to precision measurement of radiation turns out to be unexpectedly distorted, the situation is unsalvageable.

THEOREM I: <u>Radiation field in the sky</u>

An isotropic radiation field in free space is characterized by a spectrum. If this spectrum is a blackbody spectrum, the radiation field can be assigned a temperature that will work with the rest of physics. If the spectrum is not a blackbody, it has no assignable temperature.

Corollary I-1: If a temperature is assigned to a non-blackbody radiation field to replace an intensity description of the field, this temperature will not work with the rest of physics.

Corollary I-2: If a temperature is assigned to a non-blackbody radiation field to replace an intensity description of the field, it may present a view that suppresses the true extent of intensity variations.

~^~^~^~

Every single one of the above theorems was violated in one or another Big Bang satellite experiment. Behind the perimeter defense of the formidable complexity of the cutting edge space technology that warns off all but by a small group of experts, basic bread and butter physics was perverted.

It is through this stratagem that every satellite found the new things in the sky that it was looking to find for the furtherance of Big Bang.

CHAPTER III-7
Redshift concepts

III-7.1 Redshift in physics

We will discuss here only redshift of electromagnetic waves, and especially light. It is a well-established phenomenon of physics. We use the following symbols:

v_1, v_2= frequency of light at emission, reception
λ_1, λ_2 = wavelength of light at emission, reception

Consider a source of light at a location S and an observer at a location O. The two are moving with respect to each other, and the velocity v is the velocity with which S is moving away from O along the line of sight.

Let S emit a "single frequency" light at the frequency v_1 (wavelength λ_1 = c/v_1). Then this light is received by the observer at a lower frequency v_2 (wavelength λ_2 = c/v_2). The frequency has shifted downward due to the velocity v of the source. This shift in frequency towards lower frequencies (longer wavelengths) is called redshift. The frequency shifts towards 'redder' color of light. The speed of light, c, of course does not change in this process.

We discussed single frequency for this explanation. Of course the entire band of frequencies emitted by the source shifts in this way.

This redshift phenomenon is easy to understand in terms of the Doppler shift mechanism. Over the time interval one wavelength of light is emitted by the source, it has already moved a distance v/v_1= $v\lambda_1/c$. Therefore that wavelength is stretched out by the amount

$$\Delta\lambda = \lambda_2 - \lambda_1 = v\lambda_1/c.$$

It is customary to present redshift in terms of a quantity z such that

$$z = \Delta\lambda / \lambda_1 = v/c$$

This is the basic idea. When relativistic effects are included (for v approaching c) the discussion is modified.

III-7.2 Redshift in classical astronomy

Redshift in classical astronomy, as discussed in Section I-5, follows the above discussion, with the observer being an astronomer on the Earth and the source being a star or a galaxy. Of course in this case we cannot know the emitted frequency v_1. So one observes certain "marker" spectral lines of known frequencies. When the spectrum of the light arriving at the telescope shows these lines at lower frequencies, redshift for the source is deduced. From this, it is further deduced that the star is in motion at a velocity v relative to the Earth.

This interpretation of the spectral shift led to the classical view of the expanding Universe, which corroborated the then developing ideas of what would later come to be Big Bang cosmology. When this interpretation of redshift in astronomy was first introduced, no alternative physics explanations of the shift were advanced.

III-7.3 Redshift in modern Big Bang cosmology

However, Big Bang cosmology would soon change the interpretation of the observed astronomical redshift from the Doppler shift mechanism to a stretching-of-space mechanism consistent with the Big Bang mathematics. As space in the nascent Universe continually expanded, any light traveling through this space also "expanded" with it, meaning that the wavelength of the light got continually stretched out.

This process of stretching of space also modifies the inverse square law of dilution of light.

In the case of Doppler shift, the change in wavelength $\Delta\lambda$ is established as soon as the light leaves the source. In the case of stretching of space, this quantity is established over the entire travel history from the source to the observer, and thus must be expressed as a path integral.

III-7.4 Stretching of wavelength in physics

Thus the process of stretching of wavelength of light traveling through empty space has been fully accepted in physics as a valid phenomenon, in order to support the Big Bang theory.

Now we may ask: So what if such a stretching is not due to stretching of space? What if the stretching of wavelength occurs spontaneously due to some need of basic physics? If such a need can be identified, should that not be the operating physics here rather than the hypothetical stretching of space?

In Appendix D, I discuss this subject further.

This thing, what is it in itself, in its own constitution?
What is its substance and material?

Marcus Aurelius, *Meditations* Book VIII, 11

BOOK IV

THE BOOK OF MISAPPLIED KNOWLEDGE

The new breed of high-tech experimenters – from the accidental discoverers Penzias and Wilson to the fully geared up BICEP2 youth corps – has brought to Big Bang cosmology awe-inspiring technological excellence.
Not even the strongest dissidents have dared question these experiments.
The author is the first to exhume these enshrined experiments and do a thorough post mortem on them.
What emerges is scandalous, to say the least.
Young astronomers fancied themselves cutting edge engineers, and designed worthless high-tech contraptions.
Ineptness, sham, scam and fraud – these are the elements that went into the making of the phenomenal successes.
Contrary to popular belief, Big Bang cosmology has no experimental legs at all to stand on.
Contrary to popular belief, CMB is not relic radiation.
It has never been shown to have anything to do with Big Bang cosmology.

> The world clings to its old mental picture of the stock market because it's comforting; because it's so hard to draw a picture of what has replaced it; and because the few people able to draw it for you have no interest in doing so.
>
> Michael Lewis, *Flash Boys*

CHAPTER IV-1
The Penzias-Wilson experiment

IV-1.1 Language of power

To any seasoned Antenna & Microwave specialist, the discovery of Arno Penzias and Robert Wilson would appear most baffling at first sight. How can a single frequency measurement at 4 GHz discover an entire spectrum which peaks out sharply near 200 GHz – and is skewed to boot?

But let us go through the entire science of this discovery.

What did Penzias and Wilson *actually* observe? They pointed their antenna (Section II-1) to the sky and found a microwave noise power P_{obs} at ~ 4 GHz that was higher than expected. We will somewhat loosely use the concept of power addition and subtraction, for this is adequate for our purpose. Referred to the input of the receiver/preamplifier of their measuring set-up, they observed this excess spectral power:

$$\Delta P = P_{obs} - P_{expected} \tag{IV-1}$$

They also determined experimentally that this mystery excess radiation was isotropic, unpolarized and constant in time. That was the full extent of their observational findings.

The first thing that needed to be done now – as any radio astronomer would do – was to estimate from this observation the absolute 4 GHz spectral intensity I_v in the sky corresponding to this excess observed power [cf. Eq. (III-5)]:

$$\Delta P = C\, I_v\, \Delta v \int \int A_e\,(\theta,\, \varphi)\, \sin\theta\, d\theta\, d\varphi, \tag{IV-2}$$

where $A_e\,(\theta,\, \varphi)$ is the effective collecting area of the antenna. The quantity C, we remember, is the factor that takes care of any mismatch between the antenna polarization and the

143

polarization of the incoming wave. The φ (azimuth) integration is from 0 to 2π, and the θ (elevation) integration is from 0 (zenith) to $\pi/2$ (ground level), assuming that the mystery radiation is not coming from or through the ground. The antenna here is assumed to be lossless. The quantity Δv is the bandwidth of the instrument, the band assumed to be square-shaped.

This estimated I_v would be the reportable new result – a new discovery in the sky.

If now one wanted to *theorize* fancifully (without any evidence whatsoever) that this I_v comes from a presumed blackbody spectrum in the sky of intensity B_v (T_{bb}) of temperature T_{bb}, then one logically needed to calculate T_{bb} from the equation

$$I_v = B_v\ (T_{bb}) \tag{IV-3}$$

If it turned out now that this T_{bb} was close to the predicted Big Bang blackbody temperature T_{BB}, then one could *speculate* that Penzias and Wilson *may have* observed one point on the far trail of the relic radiation spectrum from the Big Bang. But this is not by any means a proof of existence of a blackbody spectrum in the sky.

The above calculation is the most straightforward and the only calculation that needed to be presented. Radio astronomy is after all about calculating intensities in the sky – first and foremost. However, the experimenters did not go anywhere near it – not then nor anytime afterwards.

Arno Penzias has been described as a radio astronomer by training, and Robert Wilson as an astronomer. So they were well familiar with standard calculations of the type above.

Was the above calculation possible to do? It is true that in the 1964-1965 time-frame computers were not as readily available. But for researchers at AT&T or Princeton it certainly was eminently possible, even if they had to process the old-fashioned punch cards. Other ways of calculating under reasonable simplifying assumption were also available.

Instead of reporting transparently that they observed an excess microwave power (a familiar physical quantity), Penzias and Wilson reported artfully that they observed an excess *Antenna Temperature* (a fictitious engineer's number.) This excess Antenna Temperature ΔT_A (~ 3.5 K) happened to be completely fortuitously in the same ballpark as T_{BB} which at the time was expected to be around a few degrees Kelvin.

Now the joint authors of the two papers reported the grand discovery of the Big Bang blackbody through this grand identification:

$$\Delta T_A \equiv T_{BB} \qquad \text{(IV-4)}$$

This is the story of the discovery

IV-1.2 Language of Antenna Temperature

But let us by all means look at the problem in the language of Antenna Temperature in which the discovery has been posited. Let us stick to the very basics and keep an open mind.

First we reproduce the equation for Antenna Temperature [Equation (III-8)]:

$$P_{noise} = kT_A =$$

$$(1/4\,\pi)\,C \int_0^{2\pi} \int_0^{\pi} k\,T(\theta, \varphi)\,p(\theta, \varphi)\,\sin\theta\ d\theta\ d\varphi \qquad \text{(IV-5)}$$

The above expression can be simplified considerably if we assume that the antenna is floating freely in space, surrounded all around by an isotropic blackbody radiation field of uniform temperature T_{bb} extending to infinite distances. In that case

$$\Delta T_A \approx C\, T_{bb}. \tag{IV-6}$$

Now, in their paper Penzias and Wilson say nothing about the polarization state of their antenna except that they found the mystery radiation unpolarized. If they found this using their Holmdel Horn antenna (and not an auxiliary antenna), then their antenna was linearly polarized. Generally speaking, the sugarscoop antenna would have a linearly polarized feed. If that is the case, then $C = \tfrac{1}{2}$. So we now have

$$T_{bb} \approx 2\,\Delta T_A. \tag{IV-7}$$

So, even under our most idealized assumption that favors the discoverers, an observation of an excess antenna temperature of 3.5K would indicate a blackbody radiation field of 7 K in the sky.

Note that in order for C to equal 1, the sky radiation would have to be collected through a dual-polarized feed and the powers in the two polarizations would then have to be added before determining the Antenna Temperature. At 4 GHz, this feed would probably be a substantial waveguide contraption, certainly not something one would forget to mention in a paper where this fact was of paramount importance. If Penzias and Wilson used such an elaborate feed structure, they surely would have said so in their paper. They would in any event have discussed the polarization factor C if they had considered it.

Now, Penzias and Wilson seem to suggest that the above idealized case (of the antenna being immersed in a uniform radiation field all around) is close to their case. Although the blackbody radiation is not entering their antenna from the lower hemisphere (ground), the antenna in any event is receiving very little radiation from that direction. So they suggest that for all practical purposes we can say that the blackbody radiation field surrounds the antenna all around.

If we accept that argument – i.e., if we bend over backward to accommodate the discovery – we reach the conclusion that Penzias and Wilson discovered a blackbody radiation field in the sky at ~ 7K.

As more reports of the measurement of the Big Bang blackbody followed that of Penzias and Wilson, the temperature of the Big Bang blackbody generally trended downward from their reported 3.5 K, eventually settling down at the value clinched by John Mather, $T_{BB} = 2.735 \pm 0.060$ K. To encompass this number, the Penzias-Wilson result of ~7 K would have to have an error bar that would in itself negate the discovery.

Thus in no way, shape or form did Penzias and Wilson observe the 4 GHz point in the Big Bang blackbody spectrum. Their 1978 Physics Nobel Prize citation

... for their discovery of cosmic microwave background radiation

left over from the Big Bang had no basis in scientific fact at the time the award was made or any time before or after that.

The strange reluctance in the relevant scientific establishment – the radio astronomers in particular – in examining the experiment in any depth is a phenomenon that would repeat again and again with subsequent Big Bang experiments.

IV-1.3 Cosmic Microwave Background vs. Big Bang blackbody

There is no question that Penzias and Wilson were the first to observe what we now know as Cosmic Microwave Background radiation (CMB) – an all-pervasive, omnipresent, isotropic and largely unpolarized radiation field predominantly observed in the centimeter, millimeter and infrared wavelength region. The identification of this field as the blackbody radiation field predicted by Big Bang cosmology is where issues arise.

Is it possible to say, at least, that Penzias and Wilson made the first crude observation of the Big Bang blackbody? Based on what we know today – in the fall of 2014 – the answer is a most definitive and emphatic no.

It is significant that neither Penzias-Wilson nor John Mather (next chapter) reported the CMB intensity, but found that CMB is a blackbody. The WMAP and the Planck satellites measured the CMB intensity directly, and found that it was vastly different from a blackbody. Unfortunately for the world and for science this actual CMB spectrum has been kept a secret to this day.

Cosmic Microwave Background and Big Bang blackbody are two entirely unrelated and unconnected things. The first is an observed and established fact. The second is bad science that has been observationally disproved and discredited.

So cutting off of the bogus linkage between CMB and Big Bang blackbody closes the book on Big Bang.

Finally, the astute Antenna & Microwave engineer who would have found the Penzias-Wilson blackbody most baffling at first sight would have been entirely vindicated in the last analysis.

CHAPTER IV-2
The John Mather blackbody spectrum

IV-2.1 A Berkeley foreshadowing

John Mather made his discovery of the relic blackbody spectrum with the FIRAS experiment on the NASA COBE satellite in the November 1989 - January 1990 timeframe, and reported it at a scientific conference in January 1990. The underlying experimental philosophy he used, however, was that developed within his Ph. D. thesis project, completed in 1974. This philosophy he gained from his thesis advisor Paul Richards. It appears that this central approach was never reevaluated science-wise and engineering-wise in the 15 intervening years. Did professional NASA engineers review this idea developed thus far within the academia? If they did, it survived largely intact.

The objective of the thesis experiment as well as the COBE-FIRAS experiment was to observe the blackbody spectrum in its substantial entirety and to define the peaked and skewed contour of that curve so well that there would be left no room for doubt that this radiation spectrum *is* present in the sky. All previous experiments failed to define this contour convincingly.

The basic idea was to make measurements at many points over the entire relevant range of frequencies, ~ 10 – 600 GHz, but especially near the peak. I repeat here for emphasis what I have already explained before. An experimenter would consider two basic approaches to the problem:

Approach A: Use a number of precision "single frequency" horn antennas – each optimized for its own central frequency, each with its own receiver/power meter – to cover the entire frequency range. Obviously, the number of such antennas that can be used is limited by practical considerations which in the end may translate to funds available.

Approach B: Use a single broadband antenna to collect the radiation, and then find a method to separate out and measure the spectral content in the collected radiation.

The first approach has the advantage of being a sure thing – with little to go wrong. The second one has the advantage of providing close-packed data points, thus faithfully defining any radiation spectrum present.

This second route was chosen for the FIRAS experiment, the route that had been followed in the Berkeley experiments of Richards, Mather and others some fifteen years ago. Mather brought with him to NASA an experimental approach that he was very familiar with and felt comfortable with. He would simply pick up his thesis research where he had left off.

IV-2.2 FIRAS precursor: The tin horn telescope

From John Mather's Ph.D. thesis it is clear that he never understood that an antenna or an antenna-like radiation collecting device was scientifically the most crucial component of the experiment. This truly was the novel "research" contribution to bring to the novel project. The rest of the project was adaptation of known technology and pursuit of established methodology. In other words, Mather the graduate student's main original research would be the method of precision sample collection from the sky for the purpose of precision quantitative measurement. That crucial component is exactly where he bungled his project in most fatal ways.

150

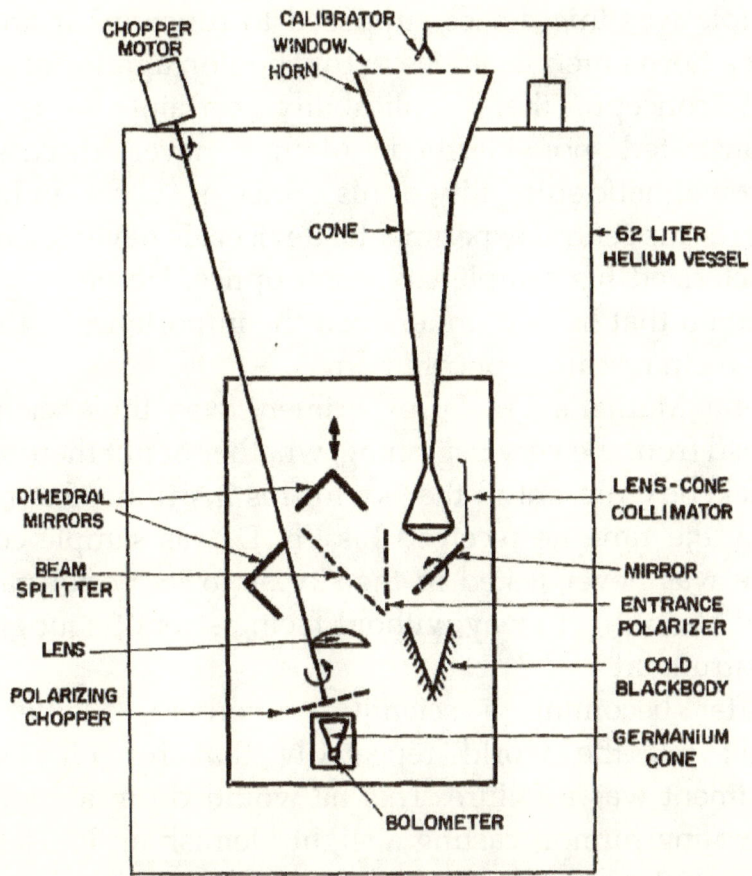

Figure IV-1: Schematic of the Ph. D. thesis experiment of John
Mather at the University of California, Berkeley (1974).
A nondescript tin horn ("CONE") was used as a radiation collector.
No scientific justification or characterization of this device – which
was the central research element of the project – was ever provided.

In his thesis experiment instrumentation (Figure IV-1),
Mather used a cone-shaped horn antenna (such as described in
Section III-1.3) for this sample collection, but used it for an
application at frequencies far above the frequency band of
operation of a horn antenna that size. So on what scientific

principle was this device supposed to function? If, instead of being a horn antenna, it was to function on a reflector antenna-based concept, that applicability certainly was neither demonstrated nor clarified. Mather never discussed the electromagnetic suitability of his choice of such a tin horn or a funnel as his central experimental device. He never scientifically characterized his sample collection optics. He never gave any indication that he had understood the importance of the optics as his main research contribution.

John Mather's Ph. D. experiment was thus scientifically doomed from the very beginning, whether or not there was a 2.7 K blackbody present in the sky for his instrument to measure. But by the time he received his Ph. D., his sample collection device was never tested in the sky as to its performance. He would cleanly get away without facing scientific judgment on his instrument.

After becoming a scientific superhero, Mather would himself tell the world repeatedly that his Ph. D. thesis experiment was a failure. This he would do in a vein of self-deprecating humor, casting a slight blemish on himself, like a beauty spot on the face of a supermodel. This is intended to enhance his image as a genius – with human frailties and all.

What was this failure? After his instrument package was launched as a balloon payload, a motor and an amplifier reportedly failed (or never functioned.) This is culpable experimenter's negligence. These components were clearly not adequately designed and tested for balloon flight (which flight cost $100,000 of taxpayers' money in 1970s dollars.) Mather had a cavalier attitude towards what he was doing, and that was what resulted in the failure.

All this is important to our discussion because all this trait would be repeated again with the NASA COBE satellite. This time the stakes would be worth hundreds of millions of then current taxpayer dollars. Moreover, NASA's reputation as a scientific research agency and Big Bang's future as a viable cosmology would ride on him.

John Cromwell Mather was given a Ph. D. by the University of California, Berkeley in January 1974 for his efforts in connection with this failed project.

Next, the Berkeley instrument package was modified and launched again by the Richards group (24 July 1974). A successful detection of the cosmic blackbody was reported in a paper submitted to *Astrophysical Journal* with a received date of 24 February 1975. This work pinned down one side of the arch-shaped spectrum, and also the peak (Figure II-2). While Mather was listed as coauthor of this paper, he had left the group by the time the paper appeared. It is not clear how much hands-on involvement he had with this project. Years later, this same experiment would be further improved by the same group (*sans* Mather), pinning down both the ascending and the descending slopes as well as the peak of the spectrum. This result was reported in 1981 (Figure II-2).

Was this seemingly successful detection of the Big Bang blackbody spectrum reported from the follow-ups of the Mather Ph. D. instrumentation valid? Today we are able to answer this question unequivocally without even referring to the experiments themselves. By properly examining the findings of the three very expensive Big Bang satellites – COBE, WMAP and Planck – we know with total certainty that such a spectrum does not exist in the sky. One must ask how a speculative theoretical prediction of an entire spectrum was repeatedly confirmed after lengthy and laborious experimentations, when nothing resembling that spectrum ever existed in the sky.

IV-2.3 Wrong-headed approach: Needless complexities

So what were the merits or defects of the said Approach B as applied to the FIRAS instrument?

In his various post-Nobel Prize lectures Mather would say that the FIRAS instrument was essentially the same as his thesis experiment, except for the External Calibrator (Figure II-5). However, there was something else that was also significantly different – although this is not apparent to non-experts.

Mather had realized late in the game the bungling he committed in his Ph. D. experiment by using a nondescript sample collection device. He needed a scientifically designed device that could be scientifically characterized and defended: a broadband antenna that would cover the desired frequency range and have precisely defined electromagnetic properties over this entire range.

Moreover, for reasons I will discuss later, the Gain and the radiation pattern of the antenna had to remain unchanged over this entire frequency range.

But such an antenna simply does not exist. Therefore, it was chosen to employ the antenna-*like* device Winston Cone (Figure III-12). This device is also called a Heat Trap because its design intent is to concentrate radiant heat. The Winston Cone never had the design intent to serve the needs of ultra-precision quantitative measurement of free space radiation. It was Mather who put it to this unique application. He modified the tin horn of his thesis to a Winston Cone.

Choosing this method meant that certain properties of conventional antennas had to be sacrificed. The first and the most crucial one is that the Winston Cone concentrates the incoming radiation from free space and produces at its output port a narrow beam of free space radiation. This beam could not be guided in a waveguide or a cable (as is the case at the focus of a conventional reflector antenna or at the throat of a conventional horn antenna), but it had to be handled differently.

Second, the issue of separating out the spectral content of this narrow beam of radiation: Here a conventional interferometric technique was used – also a continuation of the Ph. D. approach. The sky radiation spectrum and the radiation spectrum from a synthetic blackbody calibrator were subtracted from one another in the interferometer (Chapter II-2). The idea was that when the adjustable temperature of the calibrator would equal the unknown temperature of the sky blackbody radiation, the interferometer would produce a null spectrum at its output: one showing zero power at all frequencies.

This nulling technique Mather would later describe as the key to the incredible accuracy with which FIRAS reportedly clinched the blackbody spectrum.

The Winston Cone and the interferometer in this approach go hand in hand. If the Winston Cone was settled upon for some reason, then the receiver had to be an interferometer, for that is the only thing that would receive a free space beam of radiation and extract its spectral content. On the other hand, if an interferometric measurement technique was settled upon for some reason, the antenna had to be a Winston Cone, for that is the only antenna that would provide a free-space beam of radiation needed at the interferometer input.

From the various things Mather says today in his public speeches, it seems that his mind was set on the interferometer approach of his Berkeley project. It may even be that he was hooked on this in a love-of-gadgetry sense. It may also be that he felt comfortable with it, and preferred it to exposing himself to technological and engineering challenges that would be new and intractable for him.

Why was this a wrong–headed approach? First, in an epoch-making and pioneering scientific experiment calling for phenomenal measurement accuracy, one should never have given up the well-defined properties of conventional antennas in favor of a handy dandy heat trapping device. This was an inexcusable and fatal mistake to begin with.

Second, the interferometer was a very bad idea. Contrary to its advertised beauty that it could produce a very convincing null spectrum, it was never going to do so in FIRAS; not even if it had functioned ideally. Even if there were a perfect blackbody spectrum in the sky, it would be mixed in with other sources of radiation. So in actual practice the null spectrum was never going to materialize from the interferometer – that much was clear *ab initio*.

The null spectrum was a mirage around which the whole FIRAS project revolved and evolved and solidified. Furthermore, the interferometer is a device that has moving parts capable of precise nanoscale movement – yet another disadvantage for space-borne experiments. The ability to keep the input beam of radiation collimated as it travels through the interferometer is also another issue. I have not been able to find any examples of space-borne applications of such spectrometers at the time FIRAS was planned.

The external calibrator as well was actuated by motorized movement, a mechanism that proved to be problematic in ground tests, right up to the launch window.

In this way, a multiply wrong experimental philosophy born in Berkeley was pursued. Its consequences would soon become all too obvious.

It should be noted that a competing research group, that of Herbert Gush, also chose essentially the same philosophy. I do not know if one group borrowed the other's philosophy.

The discussion of FIRAS discovery should end right here, with the solid conclusion that it never happened. FIRAS was in fact a series of mistakes, each fatal in itself. Sometimes one of these mistakes here makes another mistake irrelevant or inconsequential. Logic gets a good workout here. One can stop the whole discussion upon exposing any one of these mistakes. However, to stop thus is to miss out on many more 'lessons learned', about FIRAS and about Mather and about NASA science.

And of course about an errant scientific establishment that allowed the unchecked growth of the curious John Mather phenomenon. Obvious errors and botch-ups in plain view were never noted. Obvious questions that arose were never asked. Later they would celebrate him as a superhero of science.

IV-2.4 The null spectrum approach: Foolish or diabolically clever?

FIRAS reported measuring the Big Bang relic radiation spectrum in the sky with an accuracy that would strain the credulity of anyone with any sense of such measurements. The experiment showed that there is no discernible difference between the measured spectrum and a mathematical blackbody spectrum.

Had FIRAS measured the absolute intensities in the sky and plotted them as function of frequency, and had these data points defined a mathematically perfect blackbody curve, we would have understood the discovery in a straightforward way. But in actuality FIRAS never ever measured any intensities in the sky.

So where and how does a mathematical blackbody curve enter into the FIRAS operations to provide a guiding template on which the discovery would shape itself?

What FIRAS actually recorded was the *difference* between the sky spectrum and the calibrator spectrum (Chapter II-2). When the null spectrum would obtain, then the unknown sky spectrum could be equated to the calibrator spectrum, assumed to be a mathematical blackbody. So the sky spectrum would then be concluded to be a mathematical blackbody. To summarize, this is the logical principle:

(a) A null spectrum is detected by FIRAS.

(b) Therefore, the unknown sky spectrum $\equiv$ the known calibrator spectrum.

(c) The calibrator spectrum is said to be a mathematically perfect blackbody.

(d) Therefore the sky spectrum can be plotted as a mathematically perfect blackbody.

Notice that no spectral power in the sky is ever measured directly. But has the spectral emission from the calibrator been measured to demonstrate that it is a mathematical blackbody? Never.

No spectrum of any kind – the sky spectrum or the calibrator spectrum, in sky or in ground – was ever measured by the FIRAS project. The null spectrum technique was a Trojan Horse through which to smuggle a mathematical blackbody spectrum onboard the COBE satellite. This would then become the discovery.

What the researchers did was to make some *reflectance* measurement on a surface of the calibrator material, purporting to show that it has a very low reflectivity. From that they concluded that this material is blackbody, and somewhere along the line that unproven blackbody became a mathematically perfect blackbody.

Reflectance measurements alone do not confirm a blackbody – this much is well known. The FIRAS researchers needed to perform spectral emissivity measurement with their own calibrator material near 3 K. Only *that* measurement would have told us, indirectly, what the shape of the sky spectrum is (in the event a true null spectrum had been obtained.)

So one can see that a mathematical blackbody spectrum never entered the FIRAS scheme in a scientific way. It was sneaked in through the backdoor. It then became the entire basis of the grand discovery.

In truth, with FIRAS, we had no idea what the calibrator spectrum was. Therefore we have no idea what FIRAS measured in the sky. Moreover, FIRAS was never able to pluck that much-vaunted null spectrum out of the sky.

IV-2.5 Spurious discovery of picture-perfect blackbody

It is also of importance to note that the null technique, the way it is designed, can result in spurious discovery of picture-perfect blackbody in the sky. Consider these scenarios:

(1) The calibrator does not cap the antenna (even if the instrument readings may show it does.) In this case the sky radiation will be subtracted from sky radiation, resulting in a perfect null spectrum.

(2) The calibrator does not move out (even if the instrument readings may show it has.) In this case the calibrator radiation will be subtracted from the calibrator radiation, resulting in a perfect null spectrum.

(3) The instrument somehow malfunctions, bringing very little power to the detectors. In this case a perfect null spectrum will result.

Once again, the FIRAS discussion could end here. The class could be let out right here. But we have many more instructive examples of bungling and deception to wade through.

IV-2.6 What does broadband antenna mean?

As I have explained earlier, to say that an antenna is broadband can mean many things. It is a loose term and may mean just that an antenna functions adequately over a broad range of frequencies. In a scientific measurement application, however, one must pin down what broadband means exactly. For FIRAS, this was the exact requirement:

1. The satellite-mounted antenna pattern must remain unchanged over the entire desired frequency range over the 4π solid angle.

2. The Gain of the satellite-mounted antenna must remain unchanged over this frequency range (unchanged pattern by itself does not mean unchanged Gain.)

If either of these conditions does not hold, then the antenna will itself introduce a frequency dependence (a filtering effect) on the measured spectrum. This frequency dependence will carry through to the interferometer output. It will then be ascribed to the inherent nature of the free sky radiation.

There is no indication that Mather was even aware of any of these scientific considerations, as we shall see in the following.

IV-2.7 FIRAS Antenna ground measurements

The ground measurements on the FIRAS antenna characteristics presumably had a dual purpose:

1. To provide assurances that the antenna is indeed a broadband device, having unchanged antenna pattern ("beam profile") over the entire frequency range of interest (~ 10 - 600 GHz), and unchanged Gain over this range; and

2. To serve as a back-up/standby data base in case any questions arose with the satellite in orbit.

As the second item above already implies, the tests must be done with the *satellite-mounted* antenna, and not with a free-standing unit. The data sets from the latter case is useful to the experimenter for his own analysis, but these have no bearing on the satellite-borne experiment.

Now, that sounds like a tall order. I have been to a high bay where satellites are assembled. It is neither easy nor practical to remove a satellite from its assembly area and do things with it on the antenna range. Nor is it possible to make antenna measurements in situ. However, there is perfectly acceptable alternative: Do the tests with an *electromagnetic* mock-up of the satellite. The mock-up need only model the conducting skin of the satellite. Such a mock-up can be constructed with nothing more than what is available in a hardware store like Home Depot. This mock-up than can be mounted on an improvised pedestal and its patterns measured.

This was never done with the FIRAS antenna on the ground.

To illustrate this point, Figure IV-2 shows a scheme I once used to characterize the radiation pattern of device-mounted antennas.

A second, and an equally crucial requirement is that the Gain of the antenna must be measured over the entire frequency range, using Standard Gain antennas for comparison.

This was never done with the FIRAS antenna on the ground.

Figure IV-2: A jury-rigged set-up to measure the radiation characteristics of a device-mounted antenna. The source antenna (left) and the test antenna can be connected directly to a Network Analyzer. Protractors mounted in the positioning system can measure the position angles accurately. An antenna pattern range is not needed, and the above measurements are just as accurate. The actual location of the experiment was clear of building structures. Something along this line should have been done with a mock-up of the satellite-mounted FIRAS Winston Cone.

A third, and an equally crucial requirement is that the antenna must be tested in the manner in which it would be operated on the satellite. In this case, the requirement is that the radiation at the exit port of the antenna must be measured right at that point with a bolometric detector (as would the case in the satellite application.)

This was never done with the FIRAS antenna on the ground.

A fourth, and an equally crucial requirement is that the test frequencies on the ground should convincingly span the entire frequency range of interest.

This never done with the FIRAS antenna on the ground.

A fifth, and a most desirable requirement is that some type of spectrum (not necessarily a blackbody) should have been measured with the fully assembled and staged equipment on ground. Even a flat spectrum would do. Looking at the moon from the ground might have been one option.

This was never done with the FIRAS antenna on the ground.

Finally, the mechanical actuation of the external calibrator needed to be thoroughly tested on the ground. This is one test that *was* done, and revealed enormous problems persisting right up to the launch window. Not enough time was allowed for this foreseeable problem, even with the knowledge that cooling and heating cycles that needed to occur for this test to be conducted were long. In the end the problem was thought to have been resolved. But who knows?

In essence, when the Delta rocket blasted off from Vandenberg Air Force Base with the NASA COBE satellite on board, we had a launch vehicle of the finest kind carrying a scientific instrument package that was a worthless piece of junk: a symbol of NASA's technical excellence and a symbol of NASA's scientific incompetence soared skyward in one fiery resplendence.

IV-2.8 What did they actually do on the ground?

So what exactly were the ground antenna measurements that *were* done?

Figure IV-3 shows the FIRAS antenna design and the design parameters. Two main modifications were added to the basic Winston Cone (which is the Conical Parabolic Concentrator or CPC section, CDGH.) First, a fluted "trumpet bell" section ABCHJK was added to the aperture of the Winston Cone. Its purpose was to lower the sidelobe levels and improve the broadband pattern characteristics. Second, a Conical Elliptic

Concentrator (CEC) section, EDGF, was added to the backport of the Winston Cone. Its purpose was to collimate the concentrated radiation emerging as a divergent beam from the backport of the CPC and transform it to a more or less parallel beam (coming out of the back port of the CEC.) The acceptance angle of the overall antenna was 7°.

The pilot study of this antenna occurred over a period ending in 1986 when a report was published by the authors John Mather, Marco Toral and Hamid Hemmati (hereinafter the MTH paper.) This report contains the entire sample collection proof-of-concept science of the COBE-FIRAS experiment. This central science issue would never be opened again, but a reference would be made back again and again to the MTH paper, posited as the fully established scientific proof-of-concept for FIRAS. It is thus important to appreciate that the roots of success or failure of the FIRAS design are located here in the MTH paper – regardless of anything and everything that happened subsequently.

When the FIRAS antenna was ready for pattern measurements, the researchers brought the free-standing antenna over to a Scientific-Atlanta pattern measurement range that existed then in their home institution, Goddard Space Flight Center. However, they found that range interface to the test antenna was a standard waveguide of rectangular cross-section. The size of this waveguide that they eventually used was not reported, and a query to NASA's COBE satellite Archivist did not elicit any response.

The CEC backport of their antenna that they brought to the pattern range was circular. So they added to the backport of the CEC another, smaller CPC section to reduce the circular port size to something similar to the size of the waveguide interface (Figure IV-3).

So the mismatch between the newly designed antenna system to be tested and the range interface was two-fold: transition from a circular to a rectangular pipe, and transition between dissimilar dimensions. None of it was addressed scientifically. An unknown antenna was being tested under unknown circumstances. In the MTH language:

The microwave receiver was coupled to the horn under test by a small CPC, similar to the CEC and directly in contact with it, which focused most of the radiation onto the waveguide input to the receiver. However, no special transitions were made between the guide and CPC to enhance efficiency.

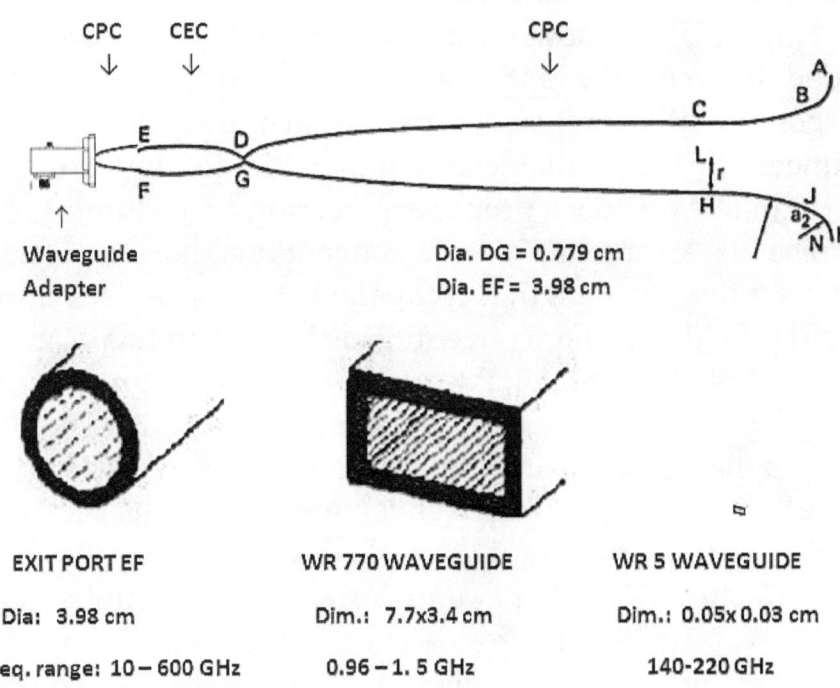

Figure IV-3: The pattern range assembly of the FIRAS antenna.

What they tried to do here is to focus the free space beam of radiation from the antenna assembly into a waveguide. This is an amateurish mistake. A waveguide – as even the name says – is a guiding structure that conducts waves in a specific frequency range. While the researchers do not say what size waveguide they used, the conundrum here can be demonstrated as follows:

Referring to Figure IV-3, one can see that if they had used a standard waveguide that is the approximate physical size of the exit port EF of the CEC, then the *frequency* of this waveguide is completely wrong. If on the other hand they had used a waveguide whose frequency range covers the peak of the cosmic blackbody spectrum (say), then the *size* of this waveguide is completely wrong – far smaller than EF.

Now, it is not known what the dimensions of the *ad hoc* CPC added between the antenna and the waveguide were. The purpose of this section seems to have been to reduce the diameter EF to a diameter comparable to the size of the waveguide. Was this a proper engineering procedure?

The FIRAS antenna is not a conventional horn antenna. For the latter the wavelength is comparable to its size. Therefore the electric field in the conventional horn antenna can flow uninterrupted to a suitable waveguide through a suitable transition – as we discussed earlier (Figure III-2). The FIRAS antenna has an aperture far larger than the wavelengths of operation. It is a Heat Trap. What emerges from its back port is a beam of free space radiation (sketched symbolically in grey bars in Figure IV-4) that is unpolarized, and "scrambled" as to the phase. The circular cross-section of the beam is many times the typical wavelength of operation. This beam can physically pass clear through a large waveguide, but it cannot be electromagnetically guided by any waveguide, the second CPC notwithstanding. Thus the mistake seems to be a result of conceptually confusing Winston Cone operation with horn antenna operation.

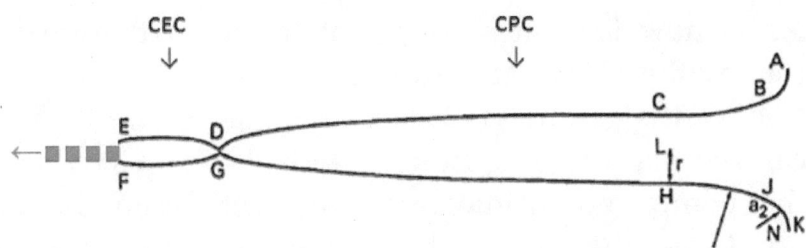

Figure IV-4: The FIRAS Sky Horn showing how a free space beam of radiation (grey dashes) emerges from its backport.

However, in the MTH set-up, some little leakage radiation would be picked up by the waveguide probe. This always happens. Thus, something resembling a pattern *would be* recorded. Such patterns would have no scientific value whatsoever.

The authors do realize to some extent and *after the fact* that they have done something wrong, and here is what they say (perhaps answering a referee) immediately following the preceding quotation, by way of an "escape clause":

Therefore, no simple statement is possible about the particular mode excitation patterns in the horn.

While this statement acknowledges the error, it does not in any way *justify* the reporting of the patterns thus acquired as scientifically valid results. With that simple and unobtrusive *mea culpa*, the researchers then proceeded to accept the faulty patterns as valid scientific results for all subsequent discussions.

This type of using of evasive language to cover up crucial

failure and then moving on as though nothing were the matter is the hallmark of John Mather's FIRAS science.

IV-2.9 Pattern measurements

Let us now look at a couple of these worthless patterns reported by the MTH paper from their range measurements as the proof of the goodness of their antenna (Figure IV-5). Here you can see that in going from 31.4 GHz to 90 GHz, the *main beam* has completely disintegrated. And yet, for an unmodified Winston Cone, the pattern should have remained completely unchanged. This may be the consequence of measuring leakage radiation which will of course behave in unpredictable ways. Other reasons are conceivable. The important point here is that there is no way to tell if these measurements reflect the properties of the FIRAS antenna assembly or are the artifacts of the measurement set-up.

Leaving aside the fact that the Winston Cone was a bad choice to begin with, how should the experiment have been done as long as one had arrived at this juncture? The NASA researchers tailored the testing of their novel antenna to the facilities of a standard antenna range they had at Goddard. What should have been done was to adapt the standard range to the needs of research: Use the positioning system of the range, but not the range receiver.

Figure IV-5 (*facing page*): Sketches of the ground measurements of the antenna beam pattern for FIRAS Sky Horn. The pattern for angles > 0 degree from the published paper was folded over to make the full beam. Hence the apparent asymmetry should be disregarded (Source: NASA).

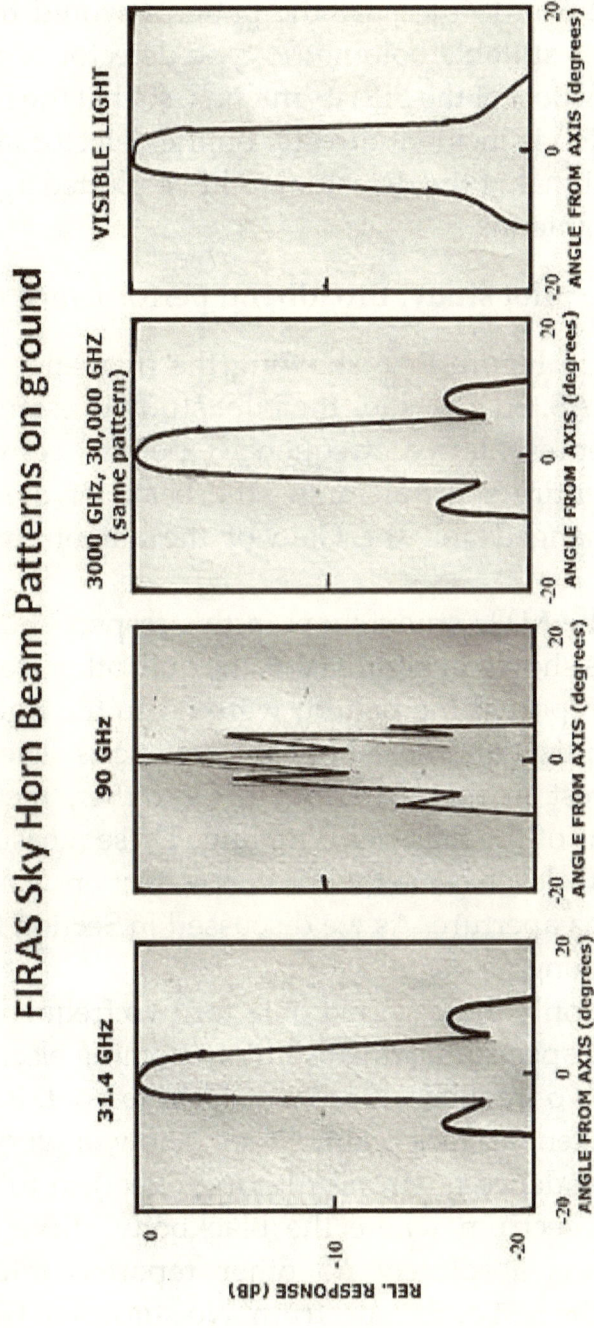

Figure IV-5

169

The correct way to measure the patterns would then have been to mount a suitable bolometric-type detector of radiation right to the back port of the FIRAS antenna, so that the free space beam of radiation is incident directly on the detector face. Then the dc output signal of the detector could be plotted against the position of the antenna.

IV-2.10 The pilot study: Broadband performance

Refer now to Figure IV-6 showing the frequency range of interest for FIRAS. So it was for the pilot study to establish that, at several representative frequency points convincingly covering this entire spectral range, the beam of the antenna remained unchanged and the Gain of the antenna remained unchanged.

What did the MTH study show in this respect? Besides the two frequencies shown in Figure IV-6, the only other frequencies at which they reported the pattern were far to the right of this diagram – at optical and near-optical frequencies. These are of no direct interest in this experiment. Or rather, these *are* of interest in terms of the litany of bungling. These measurements were made with laser beams whose cross-section was smaller than the antenna aperture. As we discussed in Section III-6, this procedure is wrong.

So we have only the reported data for two frequencies way to the left of the spectrum, and absolutely nothing else. And the data for those two frequencies are worthless to boot. But if they were to be taken at face value, they show a very strong frequency dependence (a strong filtering effect) in this crucial region of the spectrum where the blackbody curve is rising sharply. There is absolutely no other reported information developed for the rest of the spectrum. No studies of Gain were ever done.

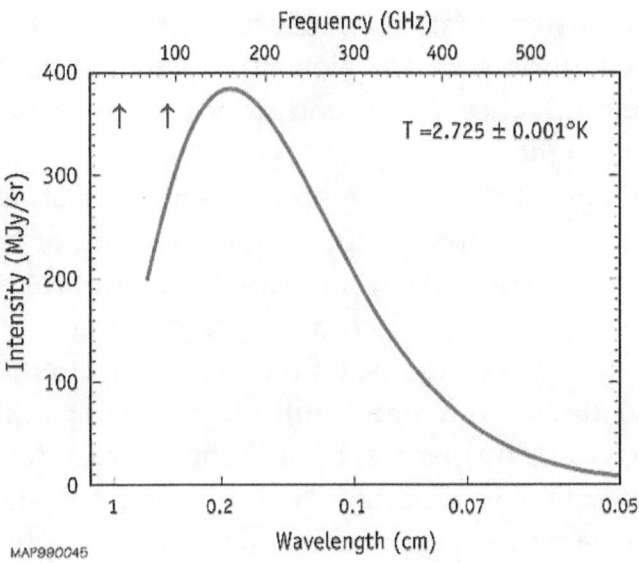

Figure IV-6: The blackbody spectrum at 2.725 K that was discovered by FIRAS, showing the frequencies (upward arrows) at which the Sky Horn antenna was tested on ground. These tests were worthless. But if taken at face value, they show strong frequency dependence. The entirety of the rest of the spectrum remained untested and uncharacterized when the instrument was launched.

Based on this study, the authors go on to report in the MTH paper that their Pilot Study is a success:

We have shown that the quasi-optical multimode horn antenna works well over an extremely broad spectral range.

How did <u>this</u> conclusion follow from <u>this</u> study?

As shocking and as unbelievable as this may sound, this is the entire scientific proof-of-concept basis of the FIRAS discovery.

In this way, the ground studies of the most crucial component of the experiment, the sample collection device, was declared successfully completed. The method of measurement

171

and the measurements themselves were faulty to the extreme. From a scientific point of view this is where all discussion of FIRAS should once again end and all the results should be written off *in toto*.

However, further developments are most instructive to follow so that one has a better understanding of how a pattern of bungling and deception continued as an unbroken chain from the beginning of the project to the present time.

In sum, when the satellite was launched, the sample collection device remained entirely uncharacterized as to its electromagnetic properties. Broadband characterization data on ground were never reported. Nor are there any statements that this these data sets were ever acquired. Only a few points at the extremities of the spectrum of interest were tested – that too with faulty measurements and with the free-standing antenna. The entire body of the spectrum remained an unknown territory as far as the electromagnetic performance of FIRAS is concerned. If the bit of test data that were on hand were to be believed, then they suggested the project should not have proceeded further. In NASA parlance, the project was a definite "No go" based on the MTH study.

In the end, with this very piece of equipment – proven to be scientifically worthless – a discovery of a spectrum with a 50 parts per million accuracy of measurement would be reported.

IV-2.11 Putting a cork in the antenna: Conceptual flaws

The FIRAS calibration technique described in Chapter II-2 was presumably designed to provide a comparison measurement scheme where the unknown sky radiation is quantified by matching it with the known calibrator radiation.

Let us examine this idea.

First, it is useful to define a couple of quantities. As we have seen before, whatever the FIRAS antenna is looking at, from its backport emerges a pencil beam of free space radiation that impinges on the interferometer input interface (a mirror.) We define the total spectral power in this beam when the antenna is looking at the open sky as $P_{v,sky}$ (v) and the same quantity when the antenna is capped by the external calibrator $P_{v,xcal}$ (v). These have the dimension of Watts per MHz, say. These are the symbolic quantities we need to examine closely.

Consider the ideal case where the sky is filled with pure blackbody radiation at the temperature T_{sky}, and the External Calibrator emits pure blackbody radiation at its temperature T_{xcal}. This temperature can be adjusted continuously so that it can match the sky. According to FIRAS operating principle, when $T_{xcal} = T_{sky}$, the following condition holds:

$$P_{v,sky} (v) = P_{v,xcal} (v) \quad \textit{for all values of } v \qquad \text{(IV-8)}$$

This condition represents the entire crux of FIRAS operating principle. It gives a null spectrum at the interferometer output. But does this null condition ever materialize?

The quantity $P_{v,sky}$ (v) under the above idealized condition depends of the pattern and the Gain of the antenna at the frequency v.

The quantity $P_{v,xcal}$ (v) has no relationship to the antenna pattern or Gain. The very concept of the pattern requires an unobstructed aperture. All scientific principles involved tell us that there is no relationship between the quantities $P_{v,sky}$ (v) and $P_{v,xcal}$ (v) whatsoever.

Have the FIRAS researchers demonstrated otherwise – either from theoretical analysis or from experimental data? No. All they have done is made a statement. In a paper published in 1999 – ten years after the discovery was made – they scientists report that

(The calibrator) fills the entire beam of the instrument and is the source of its accuracy.

What they are saying here is that the FIRAS looks at the open sky and the capping External Calibrator with exactly the same antenna pattern. Therefore, if $T_{sky} = T_{xcal}$, then $P_{v,sky}(v) = P_{v,xcal}(v)$ for all frequencies, leading to a null spectrum at the output of the interferometer.

The first point to note here is that the antenna pattern concept loses meaning when External Calibrator caps the antenna. An antenna pattern requires an unobstructed aperture to have any meaning. The statement that the calibrator fills the entire beam is foolish.

In actuality, the radiation geometry inside the antenna is completely different for the External Calibrator observation and the sky observation. The mode structures are necessarily different. There are (or can be) no studies or experiments to prove otherwise. There can be obtained no null spectra even in that idealized condition. (See also my discussion of the Herbert Gush discovery, Section IV-2.19.)

Furthermore, if the antenna pattern changes with frequency, this affects $P_{v,sky}(v)$ and $P_{v,xcal}(v)$ in different ways. So by subtracting one from the other, this effect cannot be eliminated. The filtering effect of the antenna would be fully present in the interferometer output – if that result were otherwise obtained correctly. But of course it was not.

And in actual practice in the sky, FIRAS was never able to obtain a convincing null spectrum. This was completely predictable. Anyone with any understanding of FIRAS could have predicted this. This failure was also something never disclosed to the public, letting them think that the null spectrum appeared like clockwork.

IV-2.12 The Calibration technique: A Unicorn Trap

At the time the FIRAS instrument design was developed, it was well known that the Big Bang blackbody spectrum, even if it existed in the sky, would be mixed in with other sources of radiation. The radiation the FIRAS sky antenna would see would not be a blackbody even in the ideal Big Bang sky. So the null spectrum was never going to materialize in any event. Therefore this design was known to be inapplicable from the very beginning. Why then was it adopted?

Since the sky radiation received by the antenna was never a blackbody, it was entirely immaterial whether the calibrator was a blackbody or not. There was never any need to go through the trouble of developing an ideal blackbody calibrator. This was just wasted time and money. If this technique was to be applied, the calibrator could just as well have been a flat spectrum or any other spectrum as long as it was a known spectrum.

The inclusion of a blackbody calibrator was also a most questionable decision in another respect, since it provided a blackbody template that could guide the calculations of the final result of the sky spectrum.

IV-2.13 The Apple Antenna and the Orange Antenna

But if Mather (mistakenly) believed he was conceptually correct in his calibrator design and the null spectrum approach, what he did next certainly did not support this defense.

Consider what Mather himself has reported. It was he who proudly designed and added the flared section (Figure IV-7) to the original Winston Cone design to improve the illumination of the horn significantly.

He published an entire paper dedicated to his innovation, describing how he applied the Geometric Theory of Diffraction to greatly modify the pattern of the Winston Cone with this flare section, to obtain a far better antenna. He reported:

The Sky Horn has a smoothly flared aperture to reduce diffractive sidelobes over a wide frequency range.

Thus, by his own studies, the original Winston Cone and the Cone with the flared section were two significantly different antennas, electromagnetically speaking (and also mechanically speaking.)

Now, when the calibrator swings in and corks the horn, the flare section does not take any part in controlling the illumination of the horn. It remains outside the cavity capped by the cork, and therefore "optically" inactive. When the antenna looks at sky radiation, it is the antenna with the flared section that is active. In other words, the calibrator observation and the sky observation are done with totally different antennas. So even here we see again that there can be no cancellation of the two illuminations even if both were blackbodies of the same temperature. The concept null spectrum technique has no application here.

To recapitulate: The calibrator technique was multiply flawed in ways that were foreseeable even when the technique was chosen. Each of these flaws was independently fatal.

Figure IV-7 (*facing page*): The blackbody calibrator enters the Sky Horn like a cork in a wine bottle with fluted lip. Thus a blackbody cavity is formed within the Sky Horn, and the flare section is entirely excluded from the "optics". However, the flare section is very much an active part of the optics when the Sky Horn is looking at the free sky.

176

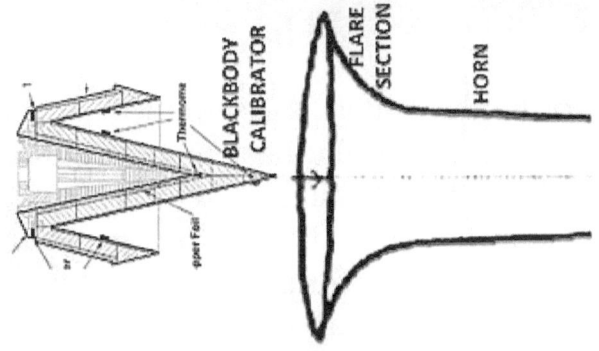

Figure IV-7

IV-2.14 FIRAS fails in the sky

When the FIRAS instrument was in orbit, it was possible to test the performance of the antenna (*satellite-mounted* this time, working with bolometric detectors this time) by sweeping its beam across the moon – which is a strong enough source of radiation for this purpose. In effect, the moon served as the source antenna of the antenna range (Figure III-4) and the positioning system of the satellite served as the range pedestal.

I have said earlier that the range measurements on ground were defective. So why should we make anything of the moon sweep pattern?

First, let us note that very little scientific and technical information has been provide for this crucial moon sweep test. Query to the FIRAS archivist did not elicit a response. So we proceed here on reasonable assumptions.

Note that even though FIRAS was receiving differential measurement data from the interferometer output, there seems to have been a way to extract the absolute power from the sky. Otherwise the moon sweep data could not have been recorded. The frequency for the moon sweep has not been disclosed. So we assume that the frequency was representative of the blackbody spectrum (otherwise the sweep test would not make any sense.)

Now, in orbit, the pattern measurement did not have the defects of the ground measurement that I outlined. While the absolute quantity of the radiation received in the in-orbit pattern measurement still should not be believed, its relative variation with angle of sweep is a faithful measurement. This is one measurement from FIRAS that can be trusted.

The results were devastating. The orbital pattern measurements are shown in Figure IV-8. The all-important flattop main beam of the Winston Cone has broken up, and shows deviations of as much as 30%. The side lobe level has come up considerably. This is the actual report:

The central portion of the beam < 3.5° is not flat topped, showing variations as much as 30%. This is only important when making measurements of sky features with angular variations smaller than 3.5°. Even then, the effect is washed out by the rotation and orbital motion of the spacecraft during a single FIRAS integration, as well as by combining multiple measurements.

It is necessary to pause. John Mather designed his entire instrument around the Winston Cone, telling us that he needed its broadband flattop beam. He thus gave up the advantages of scientifically characterizable conventional antennas. Now he is telling us that the flat top was not necessary after all.

The above example is hard evidence of fatal instrument failure that was swept under the rug, and a discovery was proclaimed – with RMS deviations of less than 50 parts per million from the peak value of the blackbody spectrum.

Figure IV-8 (*overleaf*): Orbital test of the FIRAS antenna performance was disastrous: (a) The main flattop beam broke up (left), showing deviation of as much as 30%. The frequency of this measurement could not be determined. The data for angles > 0 degree were folded over to make the full beam. (b) The sidelobes were brought up substantially over that determined on ground. The frequencies of the orbital measurements are uncertain at the source of the information, and could not be clarified.

FIRAS MOONSCAN: Optics failure

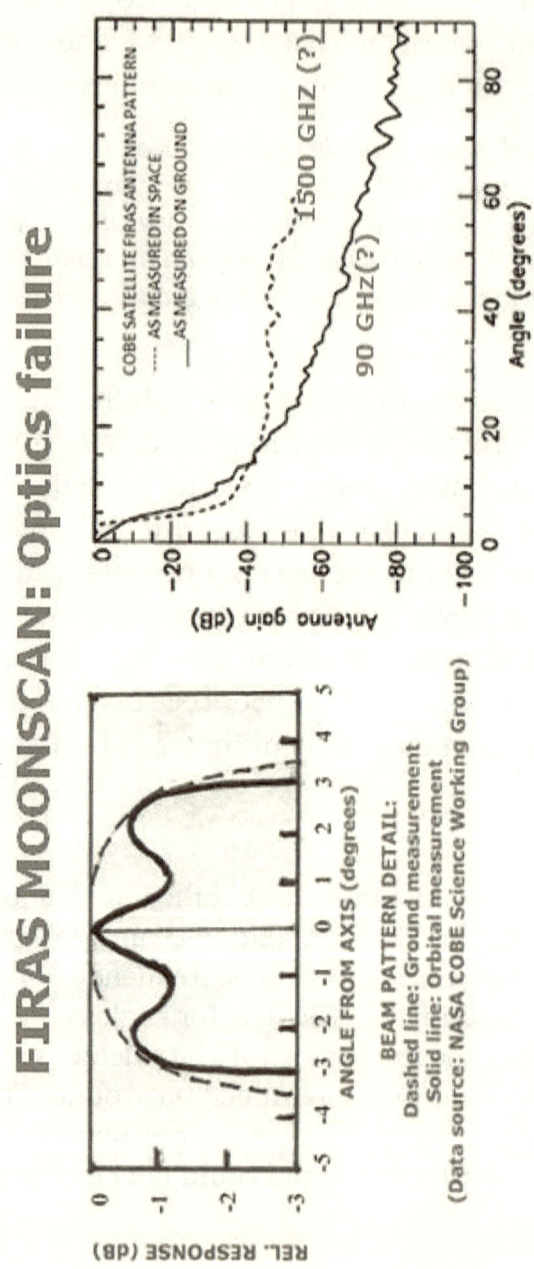

Figure IV-8: (a) Bottom figure; (b) Top figure
(Source: NASA)

IV-2.15 FIRAS post-orbit

For some ten years after the launch of the COBE satellite Mather continued to work on the satellite data. Not only that. During this period, he conducted some ground studies of his blackbody calibrator (a flight spare calibrator) to improve the accuracy of the satellite measurements.

At the end of these studies, he identified sources of errors in his instrument that needed to be removed from the data from the now defunct instrument in orbit. After the application of these various error corrections, he came up with that 50 parts per million accuracy of determination of the cosmic Big Bang blackbody spectrum.

It is necessary for the reader to ponder at some depth this aspect of the research. When developing his FIRAS antenna on the ground, Mather finalized things in great hurry, failing even to provide the minimal characterization of his instrument. After the launch was made and the instrument functioned for a while and was then turned off, and after he had already reported as successful a result as anyone might hope for, he started ground studies from scratch in great earnestness. He continued to publish papers on these studies and as he did so, the accuracy of the FIRAS measurement continued to increase apace.

With the blackbody clinched to within ±1% of the theoretical curve in 1990, one might think that the research work would have been concluded right there. The project was done. Spectacularly convincing success had been achieved. There was no need to spend time and money on gilding the beautiful lily. Not even the toughest critic could prevail in the face of this. And anyone with any hands-on knowledge of electromagnetic antenna measurements would have found even this level of accuracy to strain credulity to the extreme.

So it is an abiding mystery that another four years would be devoted on massaging the FIRAS data. Mountains of elaborate publications would follow. While Mather did not perform the essential ground tests on his instrument before launch, he would now be laboriously doing ground tests with flight spare equipment.

My own sense is that Mather knew that his 1990 spectrum was ill-gotten. He needed to get past it, to bury it and leave it behind. So he started this elaborate charade of "improving" the measurements. The new and improved spectrum he would report would supersede the 1990 spectrum and with it, all its baggage.

IV-2.16 The antenna switcheroo?

This section is necessarily speculative, and may even be altogether misconstrued. It is difficult to know. The attempt to elicit from NASA the facts of the following matter went unheeded. For this reason one is forced to speculate.

Throughout the conduct of his project – through his publications pre-launch and post-launch – Mather referred back again and again to the MTH paper whenever he needed to describe his antenna – his modified version of the Winston Cone. This consistent description had us believing that he used an antenna assembly (Figure II-6) whose cross-section – from the aperture of the antenna all the way back to the output port of the collimation section – was cylindrically symmetric. The entry aperture was a circular opening and the output port was a circular opening. When the photograph in Figure IV-9 surfaced, something altogether different surfaced.

Around January 2011 there surfaced the very first photograph of the actual free-standing FIRAS Sky Horn unit that was flown on the COBE satellite. Or at least this was the first time I saw it in spite of my vigilance of the FIRAS websites. The photograph either showed that the backport was actually rectangular, or it had an optical illusion that made it appear so.

This was a simple point to be clarified very simply. A query was sent to NASA's then COBE archivist, one Dr. Gary Hinshaw, to his official email address. It called for a yes/no answer. Hinshaw did not respond. So we will now proceed on the following assumption: Somewhere along the line Mather had some doubts about what he was doing, and he decided to radically change his antenna design without ever telling the scientific community about this.

He did not tell the scientific community when he actually made the change some time before the launch of the satellite. He did not tell them in January 1990 when he made his first public announcement of the discovery. He has not told the scientific community anything about this to this day.

So this secretive antenna switcheroo – if it happened – was yet another reason to discard the FIRAS results out of hand. But of course there had already been wave upon wave of ample reason to do so previously. The switcheroo mainly serves to tell us something about the Mather research methodology, more than anything else.

Figure IV-9 (*overleaf*): Photo of the actual FIRAS Sky Horn that went up on the COBE satellite. The exit port (bottom) appears to be rectangular. But is it? Note that the square cross-section of the leg of the workbench provides a 3-D image reference.
A simple query to the NASA COBE archivist elicited no response.

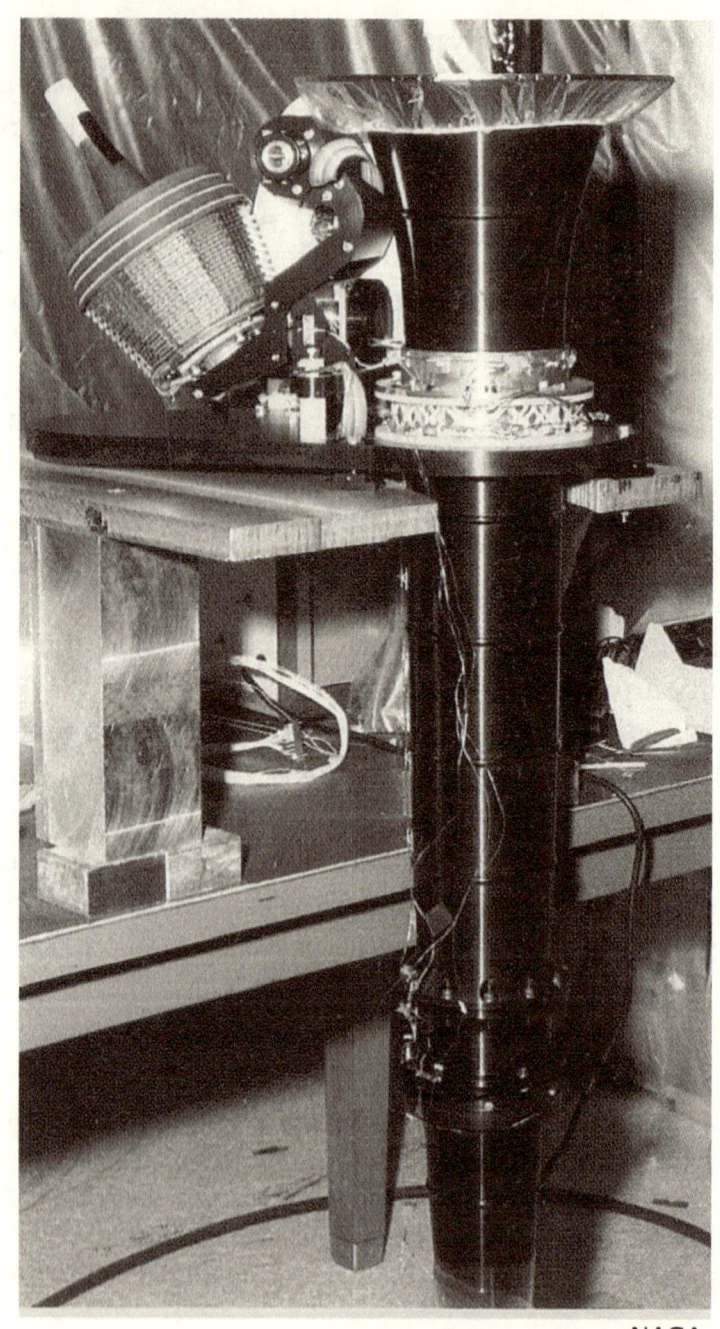

Figure IV-9

What would be the scientific problem if a switch was made?

(1) In the circular backport case, the emergent beam of radiation [power spectrum $C(v)$] has a circular cross-section. In the rectangular backport case [power spectrum $R(v)$], the beam has an oval cross-section. Therefore

$C(v) \neq R(v)$.

(2) Therefore, the output of the FIRAS reference horn (assuming it was what they said it was) – power spectrum $F(v)$ – can never cancel $R(v)$. Yet such cancellation has been reported.

$F(v) \neq R(v)$.

(3) The pattern measurements – what little was done on the original horn on the ground – no longer apply to the horn that was flown. If $G(v)$ and $P(v, \theta)$ are the Gain and the pattern, then:

$G(v)$ for the circular backport horn $\neq G(v)$ for the rectangular backport horn;

$P(v, \theta)$ for the circular backport horn $\neq P(v, \theta)$ for the rectangular backport horn.

(4) The orientation of the horn in space with respect to its axis now becomes important (e.g., with respect to the orbital pattern measurements – what plane does the pattern refer to?). But this information is not available.

So one can see that, if this NASA switcheroo did happen, it is a most serious scientific issue.

It would also say that Mather was not telling the scientific community the truth about his instrumentation. Thus, when he said he was making post-launch ground measurements on a flight spare equipment, we have no assurances that this equipment was identical to the one on the satellite. And yet, he would use these after-the-fact numbers he generated on ground to refine his satellite measurement.

IV-2.17 Was there anything salvageable from FIRAS?

As measuring instruments go, the FIRAS instrument was little more than scrap metal. Any suggestion from the Big Bang cosmologists that a critic should check Mather's numbers and calculations in order to critique his work amounts to sending the critic on a fool's errand, to neutralize him. It is completely needless to examine anything Mather has done with the numbers from the satellite. Nor is there any point to studying the mountains of publications on FIRAS. There is nothing to do here except to totally discount the measurements and the results and the publications.

Ironically, however, there *was* something salvageable in the FIRAS measurements in a negating sense.

The moon sweep experiment with a Big Bang satellite is the most ideal experiment to conclusively <u>disprove</u> the existence of the relic blackbody spectrum in the sky at a given frequency, because the experiment by its very design brackets the blackbody between the moon signal, and the system noise threshold in orbit (Figure IV-10).

In the Moon Only case, the received power is expected to mimic the antenna pattern because the moon serves as the source in the antenna pattern range. In the Moon + Blackbody case, the received power is expected to mimic the peak region of the beam and then diverge and flatten out, far above the noise floor.

186

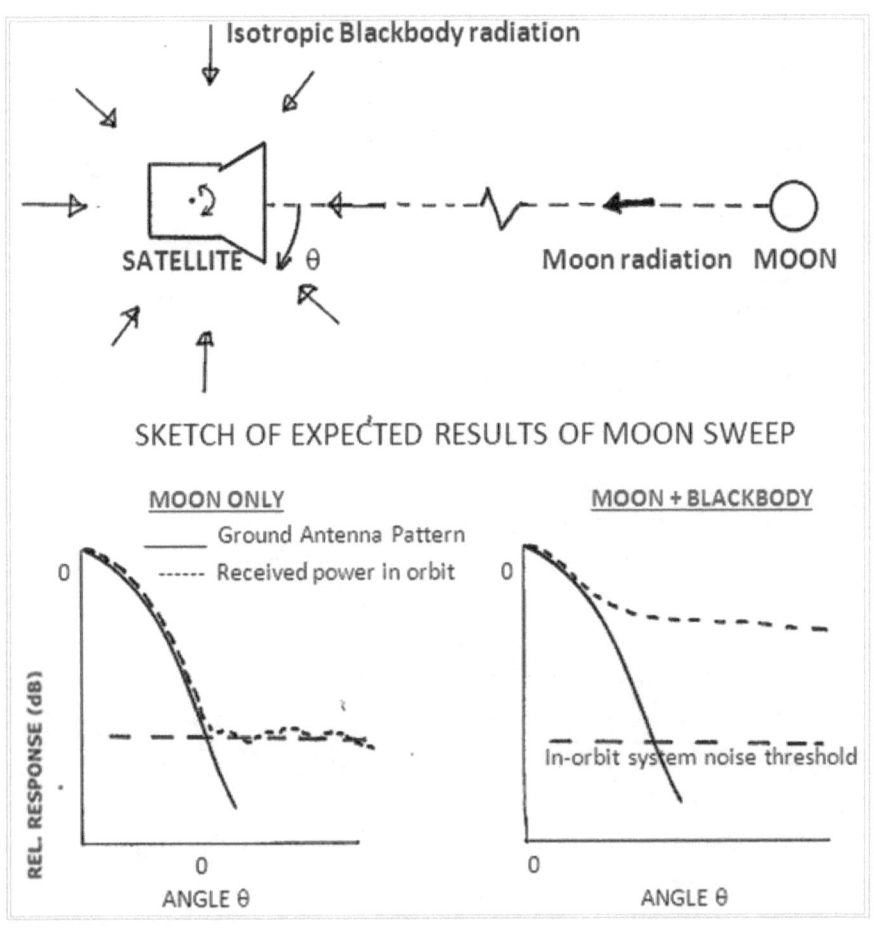

SKETCH OF EXPECTED RESULTS OF MOON SWEEP

Figure IV-10: Within the frequency range of the Big Bang blackbody spectrum, its power level detected should appear bracketed between the system noise floor and the power level from the moon.

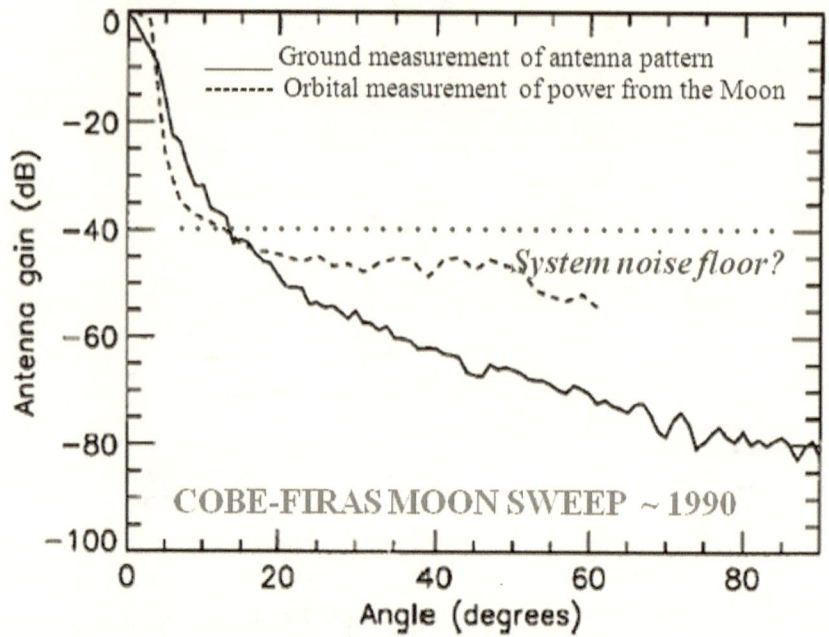

Figure IV-11: The actual result of the FIRAS Moon sweep experiment. The horizontal dotted line represents the guessed noise threshold.

Remember that these experiments were designed to show the blackbody power far above the system noise threshold. And it should be below the power levels from the moon.

So the blackbody was never seen, providing conclusive proof *from FIRAS* that it does not exist (Figure IV-11). The same conclusion would also follow even more strongly from the Smoot experiment on the COBE satellite (next chapter.) The next satellite WMAP remained questionably mum on this issue of the blackbody, but its moon sweep experiment showed clear absence of the blackbody. The next satellite Planck proved by direct and precision measurement of intensity that there is no blackbody.

IV-2.18 FIRAS New Math

If you combine (add, subtract or otherwise mix) a number with an accuracy of 50 ppm (say) with another number with an accuracy of 5% (say), what is the accuracy of the resultant number? According to FIRAS New Math, it is the better of the two starting accuracies.

The pure measurement numbers (before any processing) from FIRAS are where the final reported accuracy of measurement should flow from. So it is interesting that many times with FIRAS, these precision numbers (if that is what they were) were combined with approximate handy dandy numbers (model of galactic emission, the error estimates due to gaps in the calibrator, etc.), and the result was said to be just as precision numbers as the measurements.

This is done by convoluted tricks that involve numerical prestidigitation and linguistic jiggery pokery - cleverly folded together. Take the case of galactic emission which "contaminates" the presumed relic blackbody radiation in the sky. The final 50 ppm accuracy reported by FIRAS implies that this galactic component is known to this accuracy or better.

But the main thing to note is what they say here:

Our method for removing the Galactic spectrum assumes the Galactic emission has the fixed spectral form $g(\upsilon)$ over the entire sky. The determination of $g(v)$ is dominated by Galactic plane emission, and there could be some variation from at higher latitudes. ... Variations greater than the statistical uncertainty in any derived parameters ... would most likely be due to an inadequacy in our Galactic model.

And yet, elsewhere in the same paper, impressively tight numbers are assigned to this very effect! In the end, the numbers pass into the final result but not the cautionary language.

This is an example of the many similar data massaging procedures by which the phenomenal end accuracy of 50 ppm was got.

That is FIRAS New Math.

IV-2.19 Gush corroboration of the Mather blackbody

I have described the discovery of the Big Bang blackbody by a Canadian group of researchers from the University of British Columbia, led by Herbert Gush (Chapter II-4). Within a couple of months of John Mather's discovery, this group had independently discovered the self-same picture-perfect blackbody with equally phenomenal precision. Thus, Gush *et al* had provided an unusually strong scientific corroboration of the Mather discovery, close on its heels.

Non-expert scientists and public in general might reasonably ask: This being the case, how could the Mather discovery even be in question?

Herbert Gush's experiment was very similar to Mather's, using a differential interferometric technique and a synthetic blackbody calibrator. Gush employed two identical radiation-collecting antennas. One was capped with the blackbody calibrator. The other was open to the sky. The outputs of these two antennas were compared in an interferometer to look for a null spectrum. To achieve the null condition, the temperature of the calibrator was continuously varied. When this temperature would equal the unknown temperature of the sky blackbody, the null spectrum would result.

Herbert Gush thus discovered a blackbody in the sky with a temperature very close to that of Mather.

Clearly, the Gush group was under great pressure to come up with a positive result, given the neck-to-neck race with the NASA group. But I have not seen anything to suggest that this group engaged in any scientific transgression. It is simply that they got it wrong.

I have explained earlier that capping one antenna with a blackbody calibrator and having the other antenna look at the clear sky cannot produce the same spectra at the output of these antennas, even when the calibrator temperature is the same as the sky temperature. Gush, however, described an anecdote (without scientific evidence) that the spectra are the same.

What he did was to shroud the sky antenna with a blackbody. When the temperature of the shroud and the calibrator were the same, he found a null spectrum at his interferometer output.

This experiment is also wrong as far as its purpose is concerned. As soon as he shrouded the sky antenna, it was no longer the sky antenna. In effect he was comparing two antennas capped with blackbody of the same temperature. So naturally, a null spectrum might result, whatever the filtering effect of the antenna is. What Gush has not shown, and needed to show, is that the shrouded antenna models the free sky antenna.

Today one can look at Herbert Gush's discovery from a completely different direction and discount it out of hand. The Planck satellite has shown that the so-called Big Bang relic radiation in the sky is mixed in with copious amounts of radiation from other sources, and has to be carefully de-embedded. Gush did nothing of the kind. So how did he arrive at the picture-perfect relic blackbody?

In Section IV-2.5, I have shown how the conclusion of a perfect blackbody spectrum from the sky may result from many spurious reasons. Something of this nature may have happened, if we leave out intentional deception.

IV-2.20 The four Big Bang blackbodies of John Mather

The story of the discovery of the cosmic blackbody by John Mather played out in fascinating stages, each having its own spine-tingling, thrilling attributes. And each served an important purpose for him. For ease of discussion, I will give each of these blackbodies an appropriate name.

1. THE BERKELEY BLACKBODY (1975)

Shortly after Mather received his Ph. D. in 1974 on the basis of an experiment that failed, the continuation of that project led to the measurement of half a leg of the blackbody spectrum, with the peak clearly discerned. This result was reported by Mather and his coworkers. A few years later, the full spectrum would be reported by this group (*sans* Mather.) These results had large errors but nevertheless defined the blackbody spectrum unmistakably, and at the anticipated Big Bang relic radiation temperature.

These results (Figure II-2) no doubt served to justify the funding for the satellite quest for the selfsame blackbody, and the consequent expenditure of hundreds of millions of dollars, with Mather at the helm. The diagrams were tantalizing, as if crying out to say: Just a little better experiment, and you will see me as a picture-perfect Big Bang blackbody spectrum.

2. THE EUREKA BLACKBODY (November 1989)

It is fitting that FIRAS "discovery of all time" should have a 'Mather at the control console' story to rival Archimedes in his bathtub story. And indeed it does.

When the FIRAS instrument started acquiring its very first sky measurement data but the calibrator had not yet been actuated, the researchers devised a subterfuge. They combined the ground calibration measurements with the real sky data. And out popped there on the console a picture-perfect, pristine blackbody from the sky. If nobody exclaimed Eureka, something else equally memorable was indeed exclaimed (not mentionable here.)

This provided a most fitting anecdote for Mather's self-composed epic with himself as the protagonist. This epic, *The Very First Light*, became a commercial success and took its place in the pantheon of the great tomes of the scientific civilization such as *The Double Helix*.

3. THE STANDING OVATION BLACKBODY (January 1990)

The next stage was to present the blackbody in a little more formal setting. The calibrator was actuated and full-blown sky measurements were made. The interferograms were analyzed and the final spectrum was computed and the dense-packed data points were plotted out. It was the same telltale blackbody, fitting the mathematical Planck function within ±1%. Upon this, people were sworn to secrecy to not release this information prematurely. It needed to be handled in a manner befitting the grandest discovery of the civilization.

John Mather unveiled this graph strategically to a gathering of his scientific colleagues. They, after all, were the ones who would light the fuse to propel him to Stockholm. This was a necessary step to be fulfilled, and he did his part well.

When Mather flashed his diagram on the screen at a scientific conference in January 1990, his colleagues spontaneously broke out into a burst of standing ovation – reportedly unknown in such meetings of the distinguished and reserved luminaries. This standing ovation also has now passed into legend.

4. THE FIFTY PPM BLACKBODY (1994)

In 1994 Mather would report yet another blackbody, this time with RMS deviations from the peak value of the spectrum that are less than 50 parts per million across the spectrum. So fine was this accuracy that the error bars hid within the thickness of the pencil line of the superimposed theoretical Planck curve.

This blackbody was surely meant for that certain Swedish audience. It would stand Mather in good stead.

Every single report of these blackbodies was false. Over a period of two decades, Mather repeatedly reported observing something in the sky that was never there to observe, and that his instrument was not capable of observing.

IV-2.21 A NASA in-house design and development project

In the late seventies a rather unique decision was made within NASA: To build the COBE satellite as an in-house project. What I understand this to mean is that the work would not be contracted out to the aerospace industry, and the instrumentation would largely be designed, engineered, machined, tested, staged, and so on, all within NASA. This sounded like a bold (in a positive way) decision.

In a video produced in November of 2013, Mather explained that this step was taken because not enough information was then available to write a Request for Proposal on which the aerospace industry would bid on.

This is yet another attempt by Mather to do a snow job on people who do not have expert knowledge about how things work at the nitty-gritty level. The above argument is totally false.

While I worked for the satellite communication industry, I saw a whole range of Requests for Proposal (RFPs). They range from just a verbal description to huge documents with stringent specifications. Everything in between is possible.

For the COBE satellite, however, a great deal of information was known at the time, and it was eminently possible to put out a RFP. Such a solicitation could include the following information:

Satellite physical dimension
Planned satellite orbit
Frequency range to be covered (the spectrum)
Desired number of frequencies to cover above range
Power levels to be anticipated in the sky
Expected nature of the radiation field,

and a host of other information then available. From these the aerospace companies could develop concepts and discuss them with NASA. In the process, the project would be refined in scientific and engineering terms. More importantly, the project would be exposed to professional, practicing engineers.

None of that happened. In effect, a decision was made with great deliberation to not go down this conventional and beneficial path. Why?

My sense is that Mather steered things that way, to fit into some overall scenario he had in mind. He figured that his experience at Berkeley made him an expert on all things COBE. He wanted no challenges or alternatives to arise on his Berkeley approach. His later statements such as "an instrument builder all my life" or that his job is to do "that which has never been done before" clue us in as to what his motivation might have been in not going the industry route.

Indeed, he gradually emerged as an instrument builder to the cosmos, largely by his own description – sometimes in the form of quotable sound bites.

IV-2.22 Chapter summary

The null technique, which has been promoted as the source of the phenomenal accuracy of FIRAS, was in fact an elaborate ruse. There was never going to be any null spectrum to be obtained, even if the sought-after blackbody were present in the sky. Instead, the real purpose of the null technique was to serve as a cloak to smuggle onboard the COBE satellite a mathematical blackbody spectrum, which spectrum – through further elaborate machination – would itself become the discovery itself.

The FIRAS discovery should be adjudged dismissible outright and in its entirety in the brilliant illuminating light of three independent and converging scientific searchlight beams:

(a) No ground measurement data was ever developed showing that the FIRAS antenna was launch-worthy, a Go. What little work was done, if it was to be trusted, showed that the project was the strongest possible No Go.

(b) In-orbit tests showed that the antenna optics were a total failure.

(c) Three Big Bang satellites – COBE, WMAP and Planck – showed with total scientific clarity that the radiation strength in the sky was far below the level of the predicted Big Bang blackbody.

That is it. It is downright foolish for anyone to delve into the mountains of publications on FIRAS or to re-analyze FIRAS calculations.

The reader should ponder John Mather's two claims that are being croaked into our ears and glared into our eyes with increasing frequency from various highly visible podiums being offered Mather lately:

1. His FIRAS instrument was ideal.
2. His results were ideal because his instrument was ideal.

This would imply that going from (1) above to (2) above was an ideal path. It was anything but. Mather never obtained his vaunted null spectrum. He took years to resort to various subterfuges such as finding sources of error here and there, and torturing the data accordingly. While he was in great hurry to get the ground tests over with prior to the launch, for some ten years after the satellite launched, Mather was doing laborious ground experiments with flight spare equipment. Is this how an ideal instrument works to produce ideal results, and is this how long it takes to produce them?

To summarize: John Mather made up the 2.7 K cosmic blackbody out of whole cloth, and the 50 ppm measurement accuracy out of thin air.

The FIRAS instrument – with its calibrator swinging in and out and interferometer motors whirring – was Mather's Rube Goldberg Machine. But it was a very expensive one for the hapless taxpayers. There was nothing humorous about this.

> My personal code is, life is too long not to do what you want. It's too short not to do what you want.
>
> John Mather

CHAPTER IV-3
George Smoot and his Baby Universe

IV-3.1 An important distinction

Before we can discuss the George Smoot COBE satellite discovery, it is important that we remind or reeducate ourselves on a very important distinction that must be maintained in today's cosmology – cosmology in general, that is. This is a distinction that has been carefully erased from the public's collective consciousness by the Big Bang cosmology establishment through the use of mass psychology, with the help of the media.

The distinction concerns Cosmic Microwave Background radiation (CMB) and Big Bang blackbody relic radiation.

Cosmic Microwave Background radiation is an observed fact about the Universe – first observed by Arno A. Penzias and Robert W. Wilson in 1964. It is an all pervasive, isotropic electromagnetic radiation that seems to fill the Universe. Its spectrum and frequency domain are not known, but it is present in some strength in the frequency range represented by the Big Bang blackbody spectrum.

Big Bang blackbody radiation is a theoretical prediction from Big Bang cosmology. It is said to be the electromagnetic radiation left over from the tremendous energy released during the Big Bang explosion. It is predicted today to be a blackbody spectrum at a temperature around 3 K.

The COBE satellite is claimed to have found the CMB to have a blackbody spectrum at 2.7 K. This fact is used to say that CMB is Big Bang Blackbody relic radiation. This identity results in the conclusion that the CMB we see today is the light of the early Universe – the Baby Universe.

Furthermore, Big Bang theory also says that the Big Bang

relic radiation, while being isotropic and all-pervasive, would also have some small non-uniformity in it. If one looked in different directions, one would see small variations in the intensity of this radiation. One would see some kind of a mosaic in the sky. This is the skymap of the relic radiation, i.e., it is the map of the Baby Universe.

IV-3.2 What did Smoot actually observe?

The George Smoot experiment on the COBE satellite, the DMR experiment, has been described in Chapter II-3. His instrument looked for small variations in the intensity of the radiation as his antenna look direction swept the sky, providing full coverage of the sky. In this way he was able to collate all the differences he saw and produce a skymap of intensity variations.

Smoot made such maps in three frequencies that were representative of the blackbody spectrum.

So this much can be generally agreed on about Smoot's experiment: Smoot mapped the CMB sky (Figure II-10).

IV-3.3 What did Smoot claim to have observed?

Since John Mather's discovery established the identity between CMB and Big Bang relic radiation, Smoot claimed that he discovered the map of the Baby Universe.

With that point asserted, other results would follow: current age of the Universe, Big Bang timeline, inflation, and so on.

George Smoot was given the Nobel Prize for Physics in 2006 for mapping the Baby Universe. He was not given the Prize just for mapping CMB.

History of science offers us examples where people made discoveries but did not quite recognize the significance of what they had discovered. They then missed out on the credit.

Smoot's case is the opposite: He claimed what he had not discovered. He cannot now be said retroactively to have received the Nobel Prize for mapping the CMB sky in order to get him off the hook for the bogus Baby Universe.

IV-3.4 Why is the Baby Universe bogus?

For the simple logic that John Mather, who provided for Smoot the linkage between CMB and Big Bang, had lied. There never was a linkage. No subsequent linkages have been established. On the contrary, solid evidence has appeared from three very expensive Big Bang satellites that CMB is not a blackbody at 2.7 K, or at any other temperature for that matter.

The CMB map may have cosmological significance, but that is yet to be determined.

But before we dismiss Smoot's Baby Universe out of hand, we will address a couple of more points lest they be brought up to obfuscate the crystal clear issue I have described above.

IV-3.5 Smoot's instrument

In order to properly analyze Smoot's experiment, we need a little more understanding of his instrument than what is popular knowledge. This knowledge – as represented in Figure IV-12 (a) – has it that the instrument receives radiation from the sky through two identical horn antennas looking at different directions. The signals from the two horns are then subtracted from each other. If the horns are receiving the same signal level (i.e., the skymap has no features), then the output power is zero. But if there are features in the sky, then the output is nonzero. When this type of measurement is conducted to give full coverage of the sky, a skymap of this non-uniformity can be generated. So, in this exposition, Smoot's instrument measures only differential power between two sky directions.

In actuality, the instrument is a little more elaborate (Figure IV-12 (b). The received radiation – assumed to have random polarization – is split into two orthogonal linear polarizations, thus creating two receiver channels A and B. In each channel, the power from the two horns are combined, generating a difference and a sum of the two signals. Thus, each channel has two outputs: The Sum Signal (Σ) and the Difference Signal (Δ).

Thus we have, symbolically, for any channel:

$\Sigma = A1 + A2$
$\Delta = A1 - A2,$

where A1 and A2 refer to the signal from the two horns. Therefore:

$A1 = (\Sigma + \Delta)/2$
$A2 = (\Sigma - \Delta)/2,$

showing that Smoot was actually able to obtain the power received by each individual antenna, giving him additional information.

This means that the received power can be translated to the absolute intensity in the sky. So there should have been reported two sets of skymaps at each frequency: The Sum Channel Skymap and the Difference Channel skymap. The two sets of maps would be complementary. They needed to be studied together to arrive at any conclusion from the anisotropy.

Ideally, the Sum Channel intensity map would be a uniform sky and as such uninteresting. But we never learned what it actually is. Because it was never presented. Not by COBE nor by WMAP nor by Planck. The reason for this today is absolutely clear: The Sum Channel intensity map would have given away the crucial Big Bang secret that there is no blackbody in the sky.

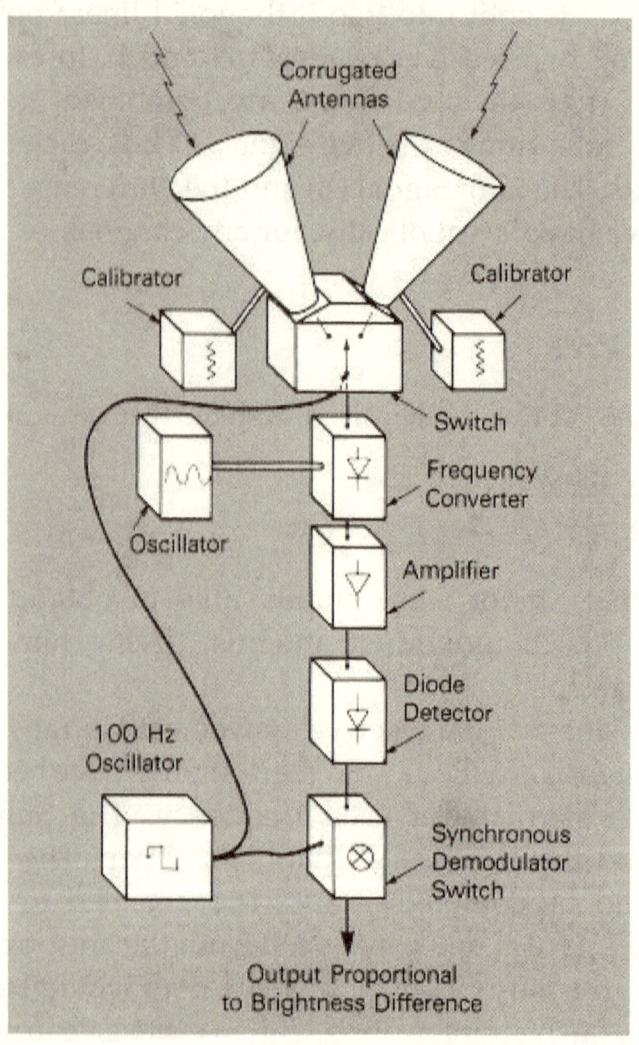

Figure IV-12: (a). Schematic of George Smoot's COBE-DMR instrument basic measurement set up.

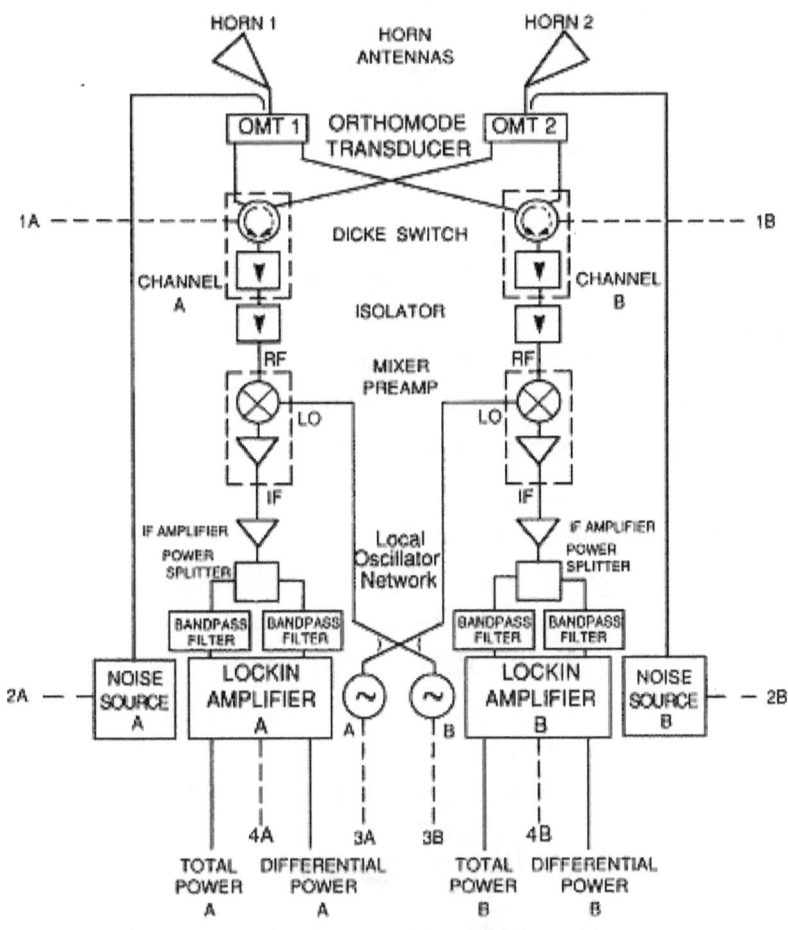

HORN 1   HORN   HORN 2
          ANTENNAS

OMT 1 ORTHOMODE OMT 2
       TRANSDUCER

1A -------   ------- 1B

       DICKE SWITCH

CHANNEL   CHANNEL
A   B

       ISOLATOR

RF   RF

MIXER
PREAMP

LO   LO

IF   IF

IF AMPLIFIER   IF AMPLIFIER
POWER   Local   POWER
SPLITTER   Oscillator   SPLITTER
           Network

BANDPASS BANDPASS   BANDPASS BANDPASS
FILTER   FILTER   FILTER   FILTER

NOISE   LOCKIN   LOCKIN   NOISE
2A — SOURCE AMPLIFIER   AMPLIFIER SOURCE — 2B
     A   A   A   B   B   B

       4A   3A   3B   4B

TOTAL   DIFFERENTIAL   TOTAL   DIFFERENTIAL
POWER   POWER   POWER   POWER
A   A   B   B

Sum(Σ) Difference(Δ) Sum (Σ) Difference (Δ)

Figure IV-12: (b). A more detailed schematic of COBE-DMR.

IV-3.6 Smoot confirmation of Mather Blackbody

George Smoot claimed (or someone in his behalf claimed) to have independently confirmed the John Mather COBE-FIRAS blackbody from Smoot's own experiment. When these three intensities were plotted alongside John Mather's data, there appeared to be a perfect corroboration (Figure IV-13).

How did they actually calculate the blackbody spectrum

from the DMR instrument? I have not been able to determine the provenance of this investigation. I queried Smoot's research group, but received no response.

We might logically assume that having obtained the power received by the Sum Channel, they converted it to the absolute intensity in the sky and then plotted all these intensities on the blackbody master diagram. But these intensities were certainly nowhere near the blackbody intensities. So where did Smoot obtain his blackbody corroboration from?

I have seen somewhere something to the effect that the spacecraft's motion through space may be used to make some conclusions in this regard.

At any rate, we see from Figure IV-13 that Smoot's three points hug the peak of the Big Bang blackbody spectrum most convincingly. One might then claim that even if John Mather's findings were totally discounted, Smoot's discovery would stand on its own - as a self-sufficient determination of both the skymap and its link to Big Bang.

We will come back to this point presently.

IV-3.7 COBE-DMR moon sweep

In Section IV-2.17 I have discussed how moon sweep experiment can confirm the presence or the absence of the blackbody spectrum in the sky. Since Smoot could measure the power received by each single antenna, he could sweep an antenna across the moon and record the received power.

Figure IV-14 shows the typical result of Smoot's moon sweep experiment. See also the blow up of the key region of this figure in Figure IV-15(a).

As with FIRAS, we see neither hide nor hair of the blackbody spectrum – not in any of the frequency channels.

Figure IV-13 *(overleaf)*: The ubiquitous Big Bang blackbody Master Diagram:
In this diagram various researchers from various parts of the world working at various times with various instrumentations have put their mark, creating a picture-perfect composite spectrum that overlays perfectly on a mathematical 2.7 K blackbody spectrum. Smoot's three data points (DMR) extend the trend towards lower frequencies after Mather's close-packed data points (FIRAS) stop.

Figure IV-14 *(page 207)*: A typical result from COBE-DMR moon sweep, showing the sweep data compared with the ground antenna pattern. Had the blackbody radiation been present, the moon data would have flattened out away from the moon to a level substantially above the system noise floor. This and other diagrams like this provide definitive proof that the blackbody is absent.

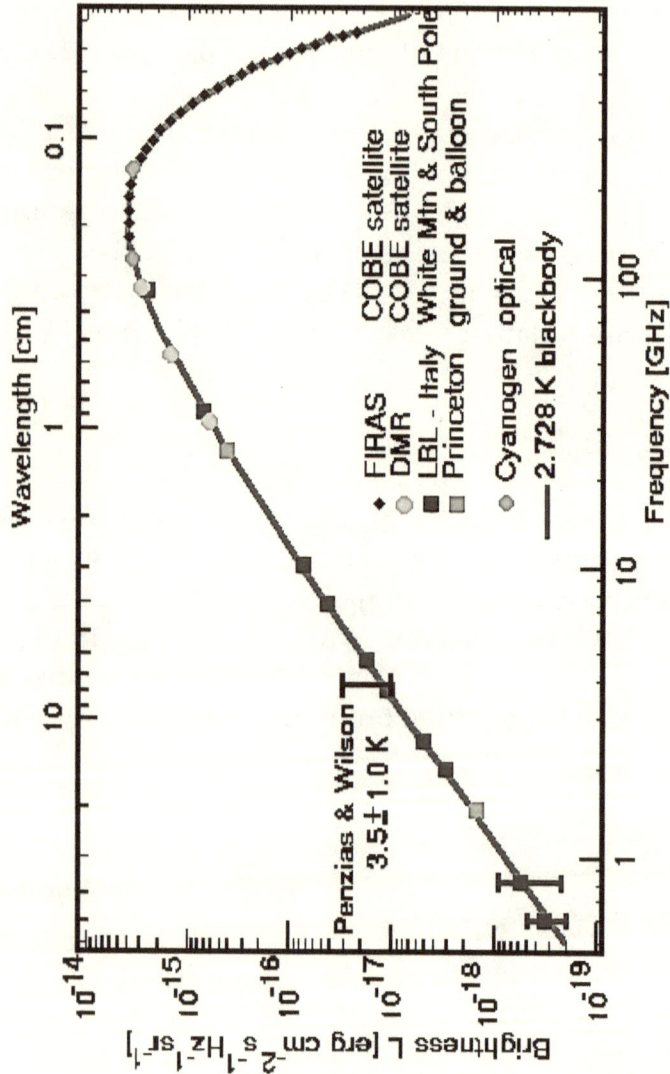

Figure IV-13

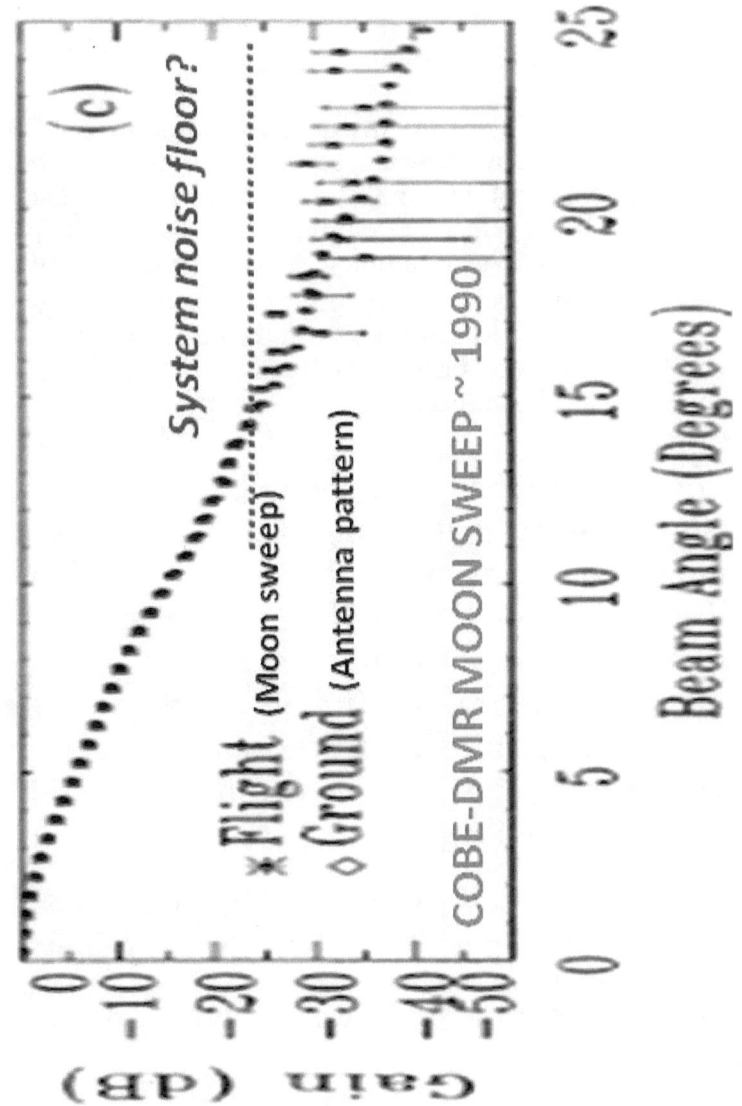

Figure IV-14

207

IV-3.8 George Smoot's unfinished Nobel discovery

Figure IV-15 shows all three moon sweep diagrams from COBE-DMR, presented by Smoot in a key paper published in 1992. Here he notices a stark problem immediately, and promises to investigate this:

> of less than 0.75% and rms deviations less than 1.5%. The flight pattern measurements become uncertain beyond ~15° from the boresight, but 99.3% of the beam solid angle is within 15° of the beam center, as computed from the ground patterns. Given the fewer assumptions, simpler analysis, and higher signal-to-noise ratio of the ground measurements, we adopt the ground measurements for both the beam patterns and solid angle values for this paper. Analysis of the flight pattern measurements will continue.

However, as far as I can find out, nothing further was said. So a big question mark hung over his proposed discovery, awaiting action from him. Awaited by whom? Certainly not the Swedish Nobel Committee. They went ahead and made their award on this unfinished job.

Yet the problem seen in this diagram was the key to everything. The picture-perfect corroboration of Figure IV-15 (a) has no corresponding information in Figure IV-15 (b). There were two clear messages:

(a) The blackbody spectrum is absent in the sky.
(b) Smoot's map does not show the Baby Universe.

George Smoot did not discover what he claimed to have discovered.

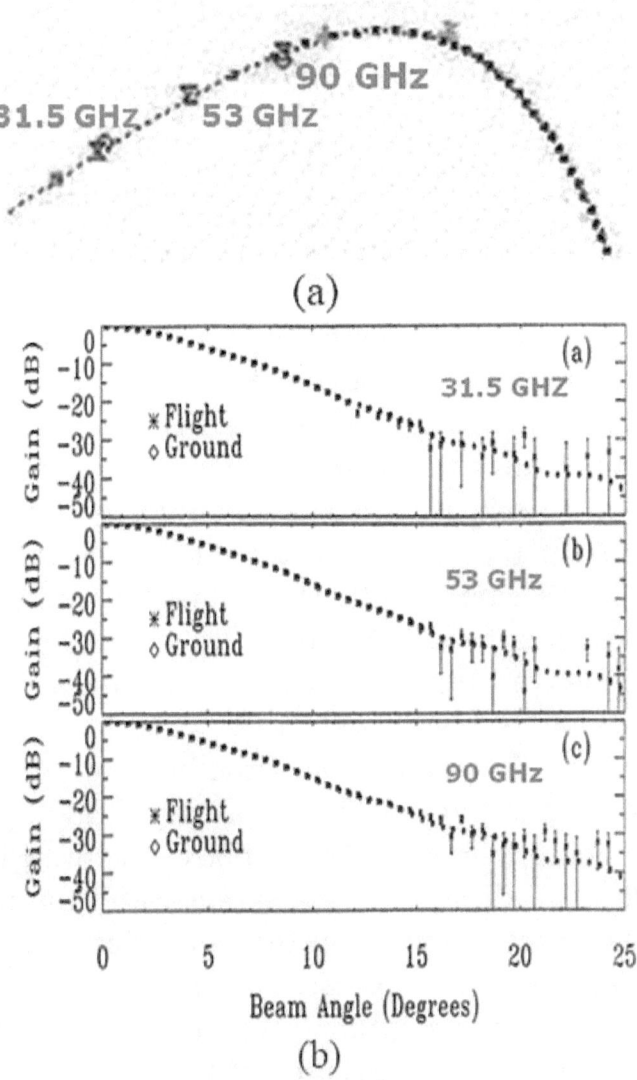

(a)

(b)

Figure IV-15: (a) Detail from Figure IV-14 showing how Smoot found the 2.7 K blackbody spectrum in the sky from the COBE-DMR experiment. His three data points (Xs with frequencies labeled) continue Mather's close-packed data points (heavy dots) towards lower frequencies in a most remarkable act of corroboration. (b) The three moon sweep experiments show no sign of that selfsame blackbody.

IV-3.9 Charles Bennett takes over

Charles L. Bennett, then of NASA, was a co-investigator on Smoot's DMR project. He was in fact a co-equal of Smoot. There is an anecdote in John C. Mather's book *The Very First Light* to the effect that there was some kind of problem with Smoot, as a result of which NASA wanted to replace Smoot by Bennett at the helm of DMR. There were also opinions that Bennett fully deserved a share of the COBE satellite Nobel Prize.

The point here is that Bennett was as much responsible and accountable for DMR as anyone else. For everything I have said in this chapter up to this point, Bennett is equally responsible with Smoot. And he was also deeply involved with FIRAS. It is widely known that Smoot had interpersonal problems with Mather. Thus it is possible that Smoot was not fully aware of the goings on within the FIRAS team. But Bennett did not have such problems. He was thus equally knowledgeable on details of the inside workings at FIRAS and DMR.

When data from the COBE satellite came in, Smoot and Bennett soon realized that there was not enough power in the sky corresponding to a 2.7 K blackbody – a power level around which they had designed their antenna and the receiver. The signal was barely above or in the noise threshold. After John Mather announced his spectacular result in 1990, it would be another two years before Smoot would announce his skymap, with its anisotropy features barely discernible.

In 1990 at the latest Smoot and Bennett knew that the blackbody was not there. The latter is not incompetent. He saw here a problem and an opportunity.

The problem – that there is no blackbody – he would have to keep mum about if he was to become the next Big Bang satellite hero. The opportunity – that he could become such a hero by vastly improving on Smoot's map by increasing the radiation collecting power – he seized on.

And this is how Bennett's WMAP satellite was eventually conceived. Bennett completely suppressed the fact that there was no blackbody, and that therefore any skymaps WMAP would obtain would have nothing to do with the Baby Universe. Instead, he maintained the CMB ≡ Blackbody charade and made his phenomenal contribution to Big Bang cosmology.

Sometimes the same fact can be put in two different ways: one upbeat and one dark. Bennett's increasing the collecting power was pitched as seeking higher resolution for the skymap. This is a correct statement. But it masked the fact that the DMR botch-up dictated higher collecting power (≡ higher resolution).

IV-3.10 WMAP moon sweep

The WMAP satellite (operating at five frequencies) also performed moon sweep experiments, and Bennett said that such data should trusted especially:

> In-flight Moon measurements are preferred to ground measurements for the following reasons:
>
> 1. The Moon is a roughly thermal microwave source, so lunar radiation allows broadband measurement of each antenna pattern. We treat the Moon as unpolarized.

It is interesting that in the case of COBE-DMR, it was said that ground data should be trusted over the moon data. For WMAP it is being said that the moon data should be trusted over ground data.

Be that as it may, the WMAP moon sweep experiment shows us again – as with the COBE-FIRAS and COBE-DMR experiments – that there is no measurable radiation in the sky above the noise threshold, other than the moon (Figure IV-16).

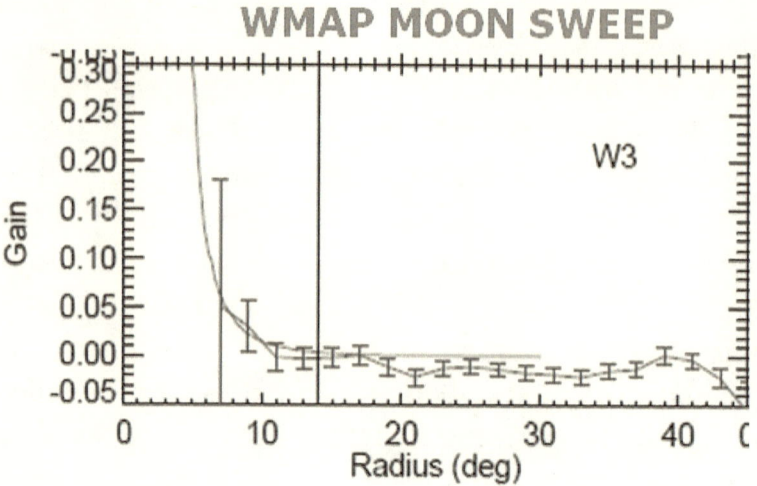

Figure IV-16: A typical WMAP diagram showing comparison between an antenna pattern (continuous line) in the sidelobe region with the moon sweep (bars), in the same angular region. This purports to show excellent agreement. It is also an excellent proof that the blackbody radiation is absent. The agreement tells us that in space the moon acts like the source in an antenna pattern range. No other radiation is present.
(The WMAP satellite was located at the Lagrangian Point L2.)

One of the first calculations Charles Bennett would have made from WMAP data is the absolute intensity in the sky at his five frequencies. In other words, in order to do all the calculations he finally reported, he had first calculated the five-frequency CMB spectrum. Because his instrument had far greater sensitivity of measurement than COBE, this was a far more reliable spectrum. But this crucial result Bennett never reported. The reason obviously was that it falsified the Big Bang blackbody as well as his own central assumption that CMB *is* the Big Bang relic radiation.

IV-3.11 Planck satellite

Planck satellite, the third in the sequence of the Big Bang satellites – provided even higher resolution than WMAP. Their skymap was thus a great improvement upon even the WMAP satellite. It has been said that the Planck measurements are as good as they can get under the present state of technology, and are unlikely to be surpassed anytime soon.

Planck satellite, positioned at the Lagrangian point L2, employed nine frequencies and calculated the absolute intensities in the sky for these frequencies. Thus they had on hand a 9-point CMB spectrum. This was the very first Planck result that had to be calculated in order to calculate anything else. Was this spectrum a blackbody? The Planck Collaboration – as if by an internal secrecy agreement – remained completely silent on this. No one disclosed this most expected result from Planck. And most surprisingly, no one asked this first and the most relevant question that needed to be asked: Where is the Planck CMB spectrum? Not the scientific establishment nor the scientific media nor the Government grant monitors.

The fact is that the Planck scientists found that the power levels in the sky were far below the 2.7 K blackbody level; that the blackbody was not there. Yet they reported results in Big Bang cosmology that presupposed the presence of this blackbody. They too continued the CMB $\equiv$ Blackbody charade.

The bottom line is this: Between WMAP and Planck, there was determined a fourteen-frequency, highly accurate CMB spectrum. The world was never told about this. This result that was thus classified by the Big Bang cosmologists was a result obtained by funds provided by American and European taxpayers for exploring the universe.

IV-3.12 Chapter conclusions

As far as a skymap of CMB goes, George Smoot was the first to show us the rudiments of this. But what he claimed was something altogether different: That it was a map of the Baby Universe. That claim was patently absurd, and wholly negated by his own experiment.

The same claim subsequently made by Bennett was far darker. Bennett <u>knew</u> that there was no blackbody from his COBE work. Afterwards he knew that there was no blackbody from his WMAP work.

The subsequent conduct of the Planck scientists in affirming the Baby Universe deceit was an historical betrayal of the scientists' creed. And also a betrayal of the taxpaying citizenry.

Since 1990, Big Bang cosmology's satellite pioneers John C. Mather, George F. Smoot and Charles L. Bennett possessed the knowledge – knowledge they developed themselves at enormous cost to the public – that there is no 2.7 K Big Bang relic blackbody in the sky. Yet to this day (end of 2014) they maintain the charade of the blackbody, causing irreparable harm to science, society and civilization, and garnering great personal benefit.

> With something like Chernobyl, the public reaction was 'Oh, my God, science has really done wrong.'
>
> George Smoot

CHAPTER IV-4
Saul Perlmutter and his Discovery Template

IV-4.1 The Discovery Template

The discovery that the expansion of the Universe is accelerating (Nobel Prize for Physics, 2011) was made by Saul Perlmutter, Brian P. Schmidt and Adam G. Riess. For economy of words, we will represent this trio of discoverers by the name Perlmutter. I described this discovery in Chapter II-5.

In order to be faithful to the discoverers' description of their discovery, I reproduce for the purpose of this discussion a diagram presented by Perlmutter in *Physics Today*, April 2003 (Figure IV-17). Here, in a redshift vs magnitude diagram, data from nearby objects (inset – solid dots) and distant supernovae (circles) are plotted. The solid dots roughly correspond to what we know as the classical Hubble redshift diagram. The circles represent the observations that led to the new discovery.

This diagram is also a "template" developed from Big Bang theory. The template is simply the diagram with all the data points removed, and presented on a transparency perhaps. The family of curves represents Universes with increasing amounts of Dark Matter (decelerating Universe), and Universes with increasing amounts of Dark Energy (accelerating Universe). Since the distant supernovae Perlmutter observed fall in the accelerating domain of this template, the discovery of the accelerating Universe follows logically.

This accelerating Universe has however some Dark Matter in addition to the Dark Energy. The exact proportion of the two has been determined and reported.

The Discovery Template initially is a single curve that is a purely theoretical construct of the distance vs velocity plot for objects expanding after Big Bang. Then, under a whole host of assumptions that are open to questioning, this curve is mapped on to a redshift vs magnitude diagram. The averring that this mapping is clinched science – as the discovery requires – is an intricate scam.

This curve is the collapsed part of the curves in the inset of the template, and is considered anchored/legitimized by classical Hubble redshift data.

The conjectures of Dark Matter and Dark Energy are then added to the theory to generate the fanning out curves: A most attractive and visually appealing discovery stratagem.

The collapsed portion of the curves is the first level scam, and the fanning out portion is the second level scam.

One of the crucial data sets that leads to the Perlmutter discovery consists of the so-called luminosity distances. These are determined using the standard candle method. There are objects identified throughout the Universe whose absolute luminosity is claimed to be known with absolute certainty. If one accepts this as established hard scientific fact, then one is led to the discovery – logically enough.

In truth the standard candle method is pure guesswork. But this is only one example of the assumptions and guesswork that must be treated as hard science in order to make the discovery.

Figure IV-17 (*facing page*): This diagram presented by Saul Perlmutter in *Physics Today* (April 2003) explains his discovery to a general science audience. It plots astronomical data on the Discovery Template, thus accomplishing the discovery.

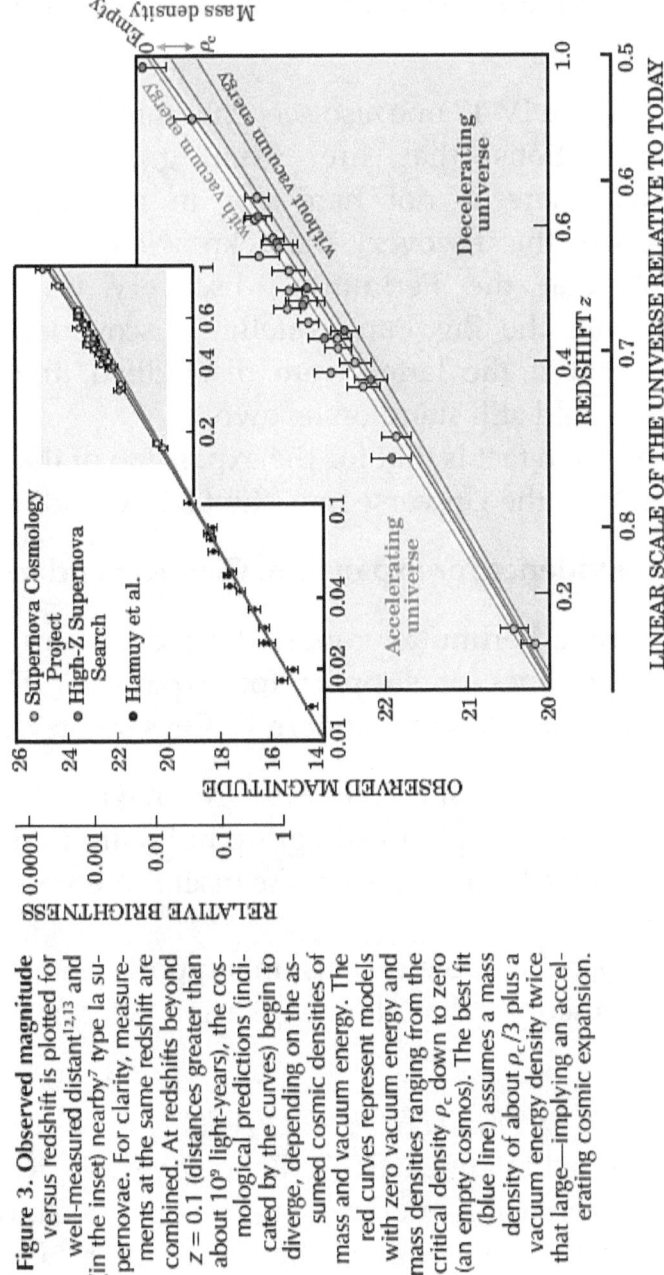

Figure 3. Observed magnitude versus redshift is plotted for well-measured distant[12,13] and (in the inset) nearby[7] type Ia supernovae. For clarity, measurements at the same redshift are combined. At redshifts beyond z = 0.1 (distances greater than about 10⁹ light-years), the cosmological predictions (indicated by the curves) begin to diverge, depending on the assumed cosmic densities of mass and vacuum energy. The red curves represent models with zero vacuum energy and mass densities ranging from the critical density ρ_c down to zero (an empty cosmos). The best fit (blue line) assumes a mass density of about $\rho_c/3$ plus a vacuum energy density twice that large—implying an accelerating cosmic expansion.

Figure IV-17

217

IV-4.2 Ground-based and satellite discoveries: Are they linked?

From Figure IV-17 one also sees that this discovery involves only observations that are from ground-based optical astronomy. There is not here nor in any other scientific expositions of the discovery the acknowledgement or even a suggestion that the Perlmutter discovery is in any way contingent on the Big Bang satellite discoveries. What this means is that if the latter were discredited, the Perlmutter discovery would still stand on its own.

But the plain fact is that for the expansion of the Universe to be accelerating, the Universe must first be expanding.

IV-4.3 Evidence for expansion: Classical and modern

At the time Perlmutter reported his discovery, there were basically two lines of support for expansion: The classical Hubble expansion and the modern COBE satellite discoveries.

CLASSICAL HUBBLE EXPANSION: I have explained this in Chapter I-3. Referring to the diagram in Figure IV-17, the black dots lying on the theory curve in the inset represents the Hubble expansion.

The classical Hubble expansion supports only one element of Big Bang theory – that of an expansion velocity proportional to distance.

MODERN SATELLITE OBSERVATIONS: The discoveries of the COBE satellite clinched Big Bang theory in its entirety in modern times, using modern science and technology. This support would later (after the Perlmutter discovery) be strengthened even more from the observations of the WMAP and the Planck satellites.

As shown in Figures IV-18 and IV-19, the entire history of Big Bang expansion was pinned down on the basis of the COBE discovery of the 2.7 K endpoint temperature (arrow in Figure IV-18 added by me.) When the Big Bang skymap was added to this, practically every idea of Big Bang theory was pinned down to extraordinary level of scientific certainty: The "explosion", the energy release, inflation, recombination, appearance of Dark Matter and Dark Energy, and so on.

And it is on the basis of this collective and synergistic picture that the Discovery Template became a scientific fiduciary. The Perlmutter Big Bang discovery has no leg to stand on if separated completely from the Big Bang satellite discoveries.

And of course if the Hubble expansion is disproved, if a physics explanation is found for the Hubble redshift, Perlmutter discovery will disappear along with the Hubble discovery and all the satellite discoveries.

Figure IV-18 (*overleaf*): Modern satellite observations have pinned down the Big Bang expansion timeline to phenomenal scientific certainty, being anchored at the 2.7 K blackbody.

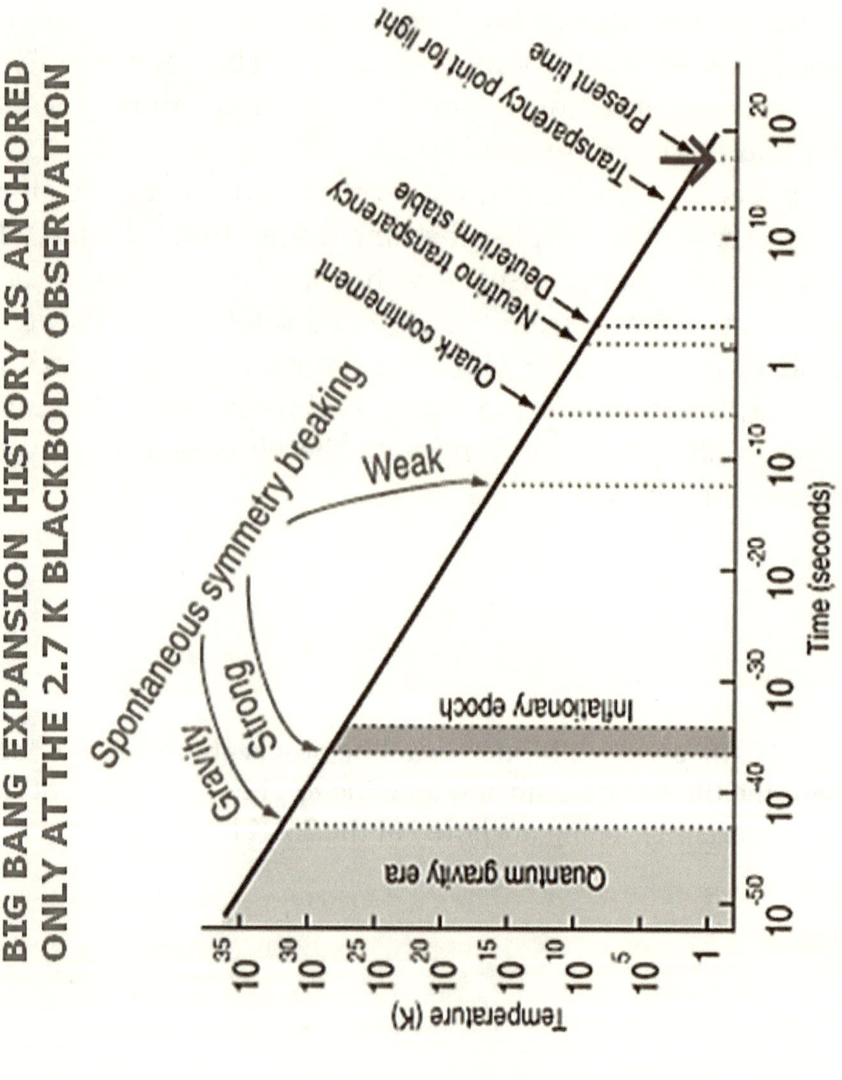
BIG BANG EXPANSION HISTORY IS ANCHORED
ONLY AT THE 2.7 K BLACKBODY OBSERVATION
Spontaneous symmetry breaking
Gravity
Strong
Weak
Quark confinement
Neutrino transparency
Deuterium stable
Transparency point for light
Present time
Quantum gravity era
Inflationary epoch
Temperature (K)
Time (seconds)

Figure IV-18

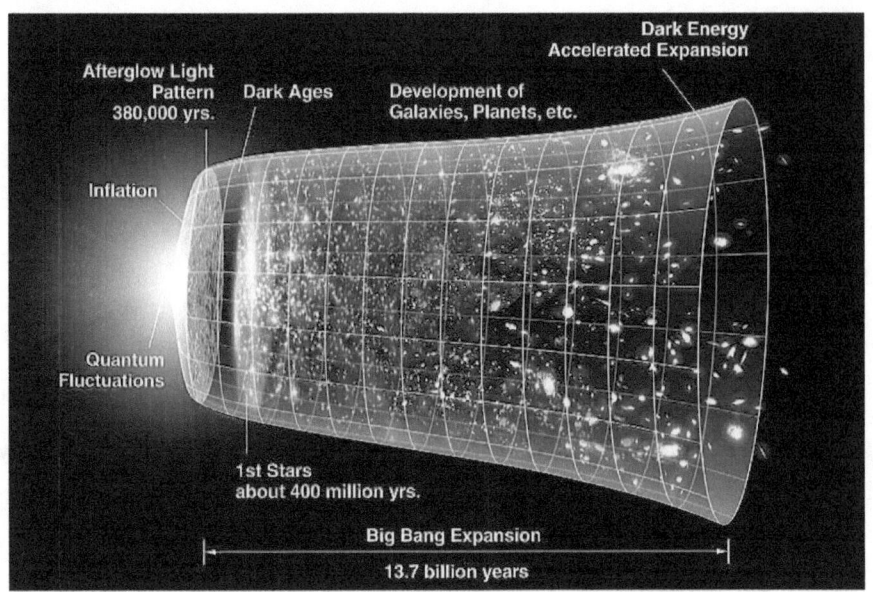

Figure IV-19: A pictorial version of Big Bang cosmology, showing how the Perlmutter acceleration fits into the timeline which would be considered entirely speculative had it not been for the satellite discoveries.

IV-4.4 Template Schlemplate

On the surface, Perlmutter's is a most convincing story as scientific discovery stories go. On a closer look it still remains so. And it remains so upon a conventional scientific review. What then is the problem here?

The problem is that we must throw out the window everything to do with satellite observation – as I have explained in the last two chapters. We must also reject any claims about the Penzias-Wilson discovery of the Big Bang blackbody for the reasons I have given in Chapter IV-1. In simple terms, the entire framework within which the Perlmutter discovery was made possible has vanished completely.

IV-4.5 Conclusion about the Perlmutter discovery

The introduction of Dark Matter and Dark Energy into the Discovery Template to create future discovery domains is rooted in the Big Bang Timeline, which timeline is clinched only by the discovery of the picture-perfect blackbody spectrum and the incredibly detailed maps of Baby Universe. That spectrum and that Baby Universe interpretation are completely bogus, and hence the timeline is also completely bogus.

Saul Perlmutter discovery rests 100% on treating his Discovery Template as a scientific fiduciary. Even if the collapsed portion of the family of curves in this template is considered legitimized by the Hubble expansion, there is no basis for using the fanning out portion as a scientific fiduciary. If the template was considered strengthened by satellite discoveries, these have all fallen apart. Saul Perlmutter does not have a discovery today, just as he did not have a discovery when he was given the Nobel Prize.

IV-4.6 Is the Hubble expansion settled science?

There has been much dissent in this regard from a handful of scientists. The dissent concerns first of all the many questionable assumptions made in mapping the observational redshift-magnitude map on to a theoretical velocity-distance map. An example of such assumptions is the way the observed magnitude of a star is used to determine the distance to the star.

Another line of dissent came principally from Halton C. Arp (who passed away in December 2013.) His observations suggested that objects claimed in Big Bang theory to be very far apart appear actually to be connected (bridged) by astronomical filaments etc.

All this dissent, however, has been rejected by the Big Bang establishment. All the Big Bang Nobel Prizes were awarded in the face of this dissent.

But beyond all this, I will elicit here considerations of foundational physics that should have arisen following the Hubble redshift observations, and that needed to be worked through by physicists before anything else took place. That discussion never happened. Instead, the subject was usurped by Georges Lemaître who hung the stretching of space explanation on it and pressed it into the service of Big Bang theory.

That cosmology explanation needs to be set aside and that physics discussion needs now to take place. It should take place within the theoretical physics community – and not within the Big Bang cosmology community.

IV-4.7 Let us have that discussion now

We now ask: As far as our knowledge of the physics of light goes, what is the basic distinction between our familiar applications and the cosmos? There is only one: Light in the cosmic setting suffers such extreme dilution as we may never have encounter in any other applications.

Let us discuss the optical region of the spectrum, and for clarity, let us discuss light of a "single" frequency v. If a star has a spectral luminosity L_v, then the intensity of this light at a distance R from the star is I_v:

$$I_v = L_v / 4\pi R^2.$$

The spectral energy density u_v at this location due to the star is

$$u_v = I_v / c,$$

c being the velocity of light.

So, for the purpose of this discussion, we say that the number N of photons (traveling) per unit volume at any point in space is

$$N = u_v \, \Delta v / hv.$$

Here Δv is a bandwidth that contains the photon energy. If we assign to each optical photon a volume of space of the order of λ^3 ($\lambda = c/v$ = the wavelength), then the number N_c of closed-packed optical photons per unit volume is

$$N_c \sim \lambda^{-3}$$

So when N falls substantially below N_c, the photons are separated in space and are not a part of the light continuum.

However, if the telescope is able to measure only continuum light incident on it, and has in fact observed the above star in this manner, what has become of the above delinking of photons?

What happens to starlight at extreme inverse square dilution that we may be dealing with in the case of the Hubble/Perlmutter redshift diagrams?

One speculative possibility I suggest in Appendix D is that the light spontaneously and continuously shifts to lower frequencies as it spreads, thus avoiding a total delinking of the photons. But for now I simply posit the above as an example of pure physics issues that needed to be looked into following the redshift observations of Hubble and others.

Three men make a tiger.

Chinese proverb

CHAPTER IV-5
The great spectral collaboration

IV-5.1 The Big Bang Master Diagram

What is the Big Bang Master Diagram? It is the result of the legendary multi-group experimental quest where independent teams – over a period of decades and working in various parts of the world – tried to construct the Big Bang blackbody spectrum in the sky. We discussed this diagram in connection with George Smoot's corroboration of John Mather's 2.7 K COBE satellite blackbody (Figure IV-13). We will discuss this diagram more fully now. Here the data points from various research groups are presented in a frequency vs brightness (intensity) diagram, with a 2.7 K mathematical blackbody curve overlain.

This must be one of the most impressive diagrams in the entire history of science. One sees very clearly that before the advent of the age of Big Bang satellites, the rising segment of the curve on the left had been completely pinned down. Then the satellite experimenters – those magnificent men with their discovery machines – took over and negotiated the peak and pinned down the falling segment.

Has anyone ever wondered how everything to do with Big Bang cosmology turns out to be uncannily correct and invariably accurate as the storyline continues to develop? The Big Bang experimenters have got a theory curve they are out to corroborate, a curve that has a sharp bend and steep slopes, and is somewhat skewed. And whoever does an experiment – however he does it, wherever he does it, whenever he does it, at whatever frequency he does it , at whatever bandwidth he uses, with whatever beamwidth he uses, with whatever angle he sweeps, whichever direction he looks in, whatever polarization he observes, whatever electromagnetic environment the observing equipment was itself in, whatever the other contributions to the observed radiation were, whatever detector he uses – puts a data point smack on that curve with pinpoint precision. Or several points. Over nearly four decades of frequency.

IV-5.2 The collaboration: A closer look

Now that I have shown what happens to the COBE satellite results of John Mather and George Smoot upon a closer look, let us go back take a closer look also at the Master Diagram.

For the Big Bang cosmologists might want to cut their losses and say: Satellite Schmatellite! Let us disassociate ourselves from the COBE satellite. We still have the data from the other researchers of the spectral collaboration to prove the existence of the 2.7 K blackbody in the sky. So Big Bang cosmology is saved harmless, as is the 2.7 K birth glow of the Universe.

Let us start from the right of the diagram and proceed leftward as we discuss the experimental data points.

Take away the COBE-FIRAS data points for reasons I discussed in Chapter IV-2.

Take away the DMR COBE-DMR data points for reasons I discussed in Chapter IV-3.

Take away the Cyanogen data. This is not a radiation measurement at all. Some very indirect arguments are made from the observation of Cyanogen line emission from the interstellar medium. When the primary measurements are in question, this type of indirect proof has no place in the discussion.

We are now left with the LBL (Lawrence Berkeley Laboratory) and the Princeton data. Table IV-1 shows a summary of some such data in a 1986 paper by N. Mandolesi *et al.* The general approach of these measurements, quantifying the sky radiation by comparing it with a blackbody calibrator, is totally wrong for the same reasons Mather's blackbody calibration is wrong (Chapter IV-2). Take all these data points away.

Why do the LBL and Princeton data all agree with one another in a ballpark sense? This may have something to do with the faulty assumption that the calibrator is emitting a 2.7 K blackbody spectrum.

Take away the Penzias and Wilson data point for the reasons I discussed in Chapter IV-1. These researchers never calculated the absolute intensity in the sky to show that it falls on the 2.7 K blackbody curve. Instead they reported having measured a sky temperature of ~ 3 K. That has been converted to a blackbody intensity and plotted in this diagram, error bar and all. That such a procedure is entertained for a single moment is most surprising.

Wavelength (cm)	Number of Observations	$T_{A,CBR}$	T_{CBR} Thermodynamic	Combined Results
12.0	6	2.48 ± 0.24	2.54 ± 0.24	2.77 ± 0.13
	18	2.81 ± 0.15	2.87 ± 0.16	
6.3	5	2.63 ± 0.21	2.73 ± 0.22	2.70 ± 0.07
	38	2.59 ± 0.07	2.70 ± 0.07	
3.0	82	2.68 ± 0.17	2.91 ± 0.17	2.75 ± 0.08
	59	2.41 ± 0.14	2.64 ± 0.14	
0.91	21	2.10 ± 0.20	2.82 ± 0.21	2.81 ± 0.12
	32	2.09 ± 0.13	2.81 ± 0.14	
0.33	29	1.00 ± 0.57	2.58 + 0.68 − 0.79	2.57 ± 0.12
	49	0.99 ± 0.09	2.57 ± 0.12	

Table IV-1: The temperatures (last column, in degrees K) of a presumed Big Bang blackbody spectrum measured by various research groups (from Mandolesi *et al*, 1986).

IV-5.3 And then there were two

Let us now take a close look at the remains of the collaboration. Figure IV-20, presented by Smoot in 1985, shows some of the data in the Master Diagram of Figure IV-13. What we see here is that the only "quality" data in this diagram that close in on the 2.7 K value come from Smoot himself. All the other data points are unconvincing. It is the Smoot dataset that clinches the slope of the rising segment of the blackbody curve on the left.

There was indeed a historical multi-group search for the blackbody. But it did not go anywhere. The much vaunted mutual corroboration among many independent groups boils down effectively to just two people who steered things towards the 2.7 K spectrum. And who are these two people? George Smoot and John Mather.

Even though Smoot and Mather both worked at Berkeley at the same time in the same scientific area and even though they were both connected to the White Mountain Research Station, they did not (it appears) know each other well then. They were brought together to plan the COBE satellite. By the time it launched, Smoot had already built the upslope (left segment) of the 2.7 K spectrum (Figure IV-20) and Mather had built the top and the downslope of the same (Figure II-2).

After these two were chosen to lead the COBE satellite project, the world did not have a prayer. If the Battle of Waterloo was won on the playing-fields of Eton, the fate of the COBE satellite was sealed in the hallways of Berkeley.

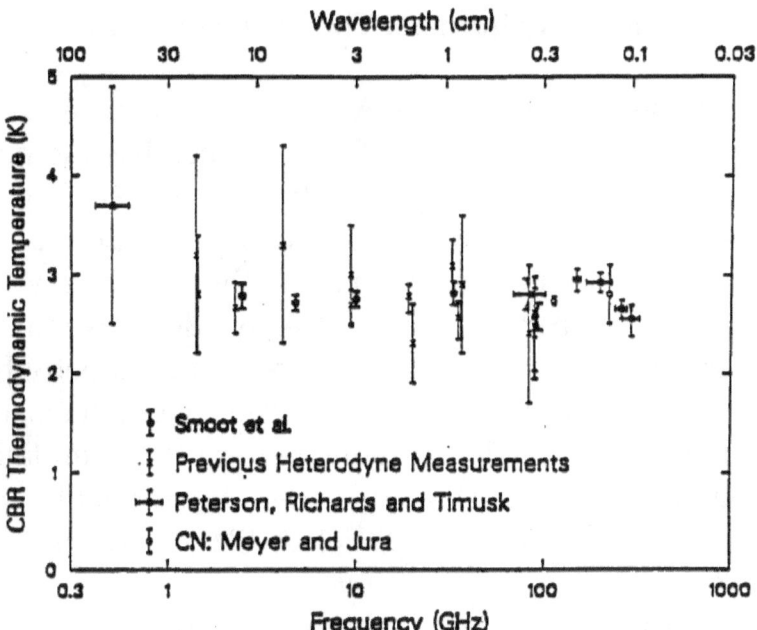

Figure IV-20: This diagram presented by Smoot in 1985 shows how, of all the measurements in Table IV-1, the only convincing measurements of the 2.7 K blackbody originate from Smoot.

FREQUENCIES OF SOME BIG BANG EXPERIMENTS

COBE Satellite - FIRAS: Spectrum
COBE Satellite - DMR: 31.5 GHz, 53 GHz, 90 GHz
WMAP Satellite: 23 GHz, 33 GHz, 41 GHz, 61 GHz, 94 GHz
PLANCK Satellite: 30 GHz, 40 GHz, 70 GHz, 100 GHz,
   143 GHz, 217 GHz, 353 GHz, 545 GHz, 857 GHz
EBEX Balloon-borne Telescope: 150 GHz, 250 GHz,
   410 GHz
SPIDER Balloon-borne Telescope: 100 GHz, 150 GHz,
   220 GHZ
BICEP Ground-based Telescope: 100 GHz, 150 GHz
BICEP2 Ground-based Telescope: 150 GHz
BICEP3: 95 GHz
KECK ARRAY Ground-based Telescope: 100 GHz, 150 GHz
POLARBEAR Ground-based Telescope: 150 GHz, 220 GHz

Table IV-2: The tremendous investment in the bogus blackbody spectrum shown in Figure IV-20.

IV-5.4 Blackbody hunters: The next generation

But this story does not end. A new generation has taken over the great quest (Table IV-2). The hunt has been joined by a number of groups. They are hunting in the sky for certain minutest of minute polarization signatures in that certain faintest of faint blackbody which is not there in the sky.

This new generation has already demonstrated great entrepreneurial spirit. Whereas the sport of blackbody-hunting was previously supported entirely by the governments, the new generation has succeeded in enlisting private foundations as their source of support. This is no small achievement.

The projects EBEX through POLARBEAR are all dedicated to this cause.

When I saw the obligatory photographs of these polarization hunters in their regulation red attire against the white icy expanse of the South Pole, this thought came to mind: If red parka and snow were all, Santa might discover.

> The hunter who always comes home with meat is a thief.
> Bantu proverb

CHAPTER IV-6
The BICEP2 botch-up

IV-6.1 What the BICEP2 project is really about

Everyone familiar with the BICEP2 Project knows that it is directed at studying the signature of the inflation era in the form of B-mode polarization swirls in the sky, created by gravitational waves produced in that era. These swirls are expected to be seen in the Cosmic Microwave Background (CMB) radiation.

Nobody familiar with the project talks anymore about the 2.7 K Big Bang blackbody radiation, which is what the BICEP2 project is really about. It is about looking for the swirls predicted to be imprinted on the said blackbody. It is about looking for something in the sky that has repeatedly been proved to be not there in the sky. BICEP2 is very much a blackbody project, like COBE, WMAP and Planck satellites.

But CMB is not that blackbody. Thus, even to begin with and even at the level of inception, BICEP2 is a misguided venture. It should never have got started, just as COBE-FIRAS should never have got started. The BICEP2 project not only got started, but quickly proliferated. Teams upon teams descended on the scene the way they do when the deer-hunting season opens.

IV-6.2 Aperture fault: Misunderstanding of telescope science

BICEP2 is a refractor telescope for microwave radiation, directed at observing the polarization characteristics of the CMB at 150 GHz (wavelength 2 mm.) It has an aperture diameter of 26 cm. Its focal plane is populated by a large number (~ 500) of small (physically and electrically) antennas with wide beams. The linearly polarized slot antennas are alternately at 90 degrees to one another to receive two orthogonal polarizations.

The aperture of the BICEP2 telescope has two fatal design faults stemming from:

A. *MISUNDERSTANDINGS* (of telescope science);
B. *MISINFORMATION* (about the amount of CMB power available in the sky.)

Personally, I would not have gone with refractor optics for millimeter wave astronomy. But this is not a criticism. The following is: While I have not sensed any consternation in the scientific establishment over this, I myself was most baffled that such a small diameter (about the diameter of a standard dinner plate) telescope could not only receive healthy signal from the dilute CMB radiation, but also produce high resolution maps of features buried deep in a very minute fraction of this radiation.

Remember that the scientists and engineers in the past went to larger and larger collecting aperture from COBE to WMAP to Planck satellites – all directed at observing CMB and at mapping features in very small components of it. The COBE-FIRAS instrument – though not directed at imaging – had a collecting area comparable to BICEP2, and was a total failure. The design of the BICEP2 telescope went in the face of all these.

However, the BICEP2 team seems to be aware of this unconventional aspect, and has offered the following scientific justification for the "novel" use of small aperture:

Small telescopes have an overlooked capability to gather a lot of light with a wide field of view. ... it was a novel approach in CMB measurements and gave us an enormous 20 degree field of view. In fact the light gathering power of BICEP is not so different from that of the 10-meter telescope looming over us at the South Pole, but BICEP's aperture is just 26 centimeters.

While it is not spelled out, this comment refers to an isotropic radiation field (which is what CMB is) where the spectral radiation flux F_v (watts per sq. meter, say, integrated over the telescope bandwidth) is the same in all directions. Let

d = diameter of BICEP2 telescope (= 26 cm);
D = diameter of 10-meter South Pole telescope (=1000 cm);
$A_{bicep2} = \pi\, d^2/4$, aperture of BICEP2 telescope;
$A_{sp} = \pi\, D^2/4$, aperture of South Pole telescope.

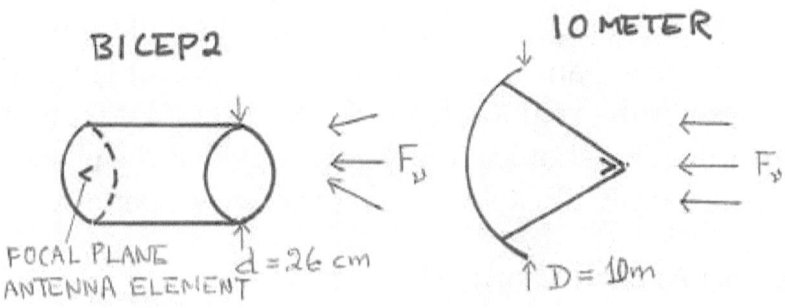

Figure IV-21: Comparison of small aperture and large aperture telescopes in terms of the capability of mapping the sky.

Referring to Figure IV-21, we see that for BICEP2 telescope, radiation enters the antenna aperture from all directions within its 20-degree field of view. For the narrow beam South Pole telescope, radiation enters only in the direction parallel to the telescope axis. This is how it is possible for the same amount of radiation (watts) to *enter the aperture* of a small telescope and a large telescope. To this extent, the statement from the BICEP2 team is correct. But this comparison is irrelevant.

What is relevant is how much radiation P (watts) is received by a single focal plane antenna element. This amount is

$$P_{bicep2} = F_v \, A_{bicep2}$$
$$P_{sp} = F_v \, A_{sp}$$
$$P_{sp} / P_{bicep2} \sim 1500.$$

So everything is in order. An antenna element at the focal plane of a small telescope receives little power, and the same at the focal plane of a large telescope receives great power. This is true in a directed radiation field or an isotropic radiation field. There is no novel way to defeat this basic physics.

The above comments pertain to the *design* of BICEP2 telescope. Is there any *observational affirmation* that the design is defective? It should be noted that Planck satellite, with vastly larger aperture than BICEP2, reports low power levels difficult to process in the region of the sky that BICEP2 mapped with such clarity. This should rightly be accepted as the observational evidence that BICEP2 telescope results do not pertain to CMB.

IV-6.3 Aperture fault: Misinformation

To design a telescope aperture for an application, one needs a design value for the energy flux available for that application. In this case it is the CMB flux F_v in the sky.

It is a matter of record that BICEP2 telescope was designed for a value of F_v corresponding to the $\sim$ 3 K cosmic blackbody discovered by the COBE satellite (Table IV-3).

However, it is an open secret within the cognoscenti that this blackbody does not exist in the sky and that the actual value of F_v may be as much as two orders of magnitude lower. This information has been available to the BICEP2 team for many years. But the academics have such an inflated collective ego that they would rather design wrong experiments and report wrong results than to acknowledge that a grave science fraud was committed with their COBE discovery.

The above two faults in BICEP2 design constitute more than enough reason to retract the BICEP2 discovery. However, in the following section I will discuss the BICEP2 imaging technique and associated faults for good measure. I will describe how a modern high precision multi-million dollar scientific instrument is periodically given a whirl the way Buddhist prayer wheels are periodically given a whirl.

TABLE 1
MODELED DETECTOR LOADING FROM ELEMENTS IN THE OPTICAL PATH

Element	T_e [K]	Emissivity	Loading [pW]	T_{RJ} [K]
CMB	3	1.00	0.12	
Atmosphere	230	0.03	2.0	
Upper Forebaffle	230	1.00	0.65	
Window	230	0.02	1.0	
IR Blocker 1	100	0.02	0.45	
IR Blocker 2	40	0.02	0.18	
IR Blocker 3	6	0.02	0.01	
Lenses	6	0.10	0.07	
Total			4.5	21

Harvard University

Table IV-3: BICEP2 Collaboration's estimate of signals from various sources. The CMB is taken to be a 3 K blackbody, contributing 0.12 picowatt to the overall power budget. In reality this signal may be 10-100 times smaller, and so much smaller than all the other masking sources – by the BICEP2 team's own reckoning.

IV-6.4 Focal plane fault

Refer to Figures IV-22 and IV-23 illustrating the BICEP2 telescope optics and imaging technique. The BICEP2 focal plane is populated with a two-dimensional array of slot antennas cut in a metal plane, in a rectangular grid arrangement. There are 512 such slots in the focal plane, and an equal number of detectors. The slots are alternately "horizontal" (parallel to the x axis, say) and "vertical" (parallel to the y axis) – to receive two orthogonal polarizations.

Figure IV-22 shows details of a portion of the BICEP2 focal plane, looking up at it from the bottom. You can see the horizontal and the vertical slots and the associated microstrip circuitry (connecting the antennas to the detector components - not shown.) While the slots and the circuitry are separated by a dielectric layer, it seems to be transparent, thus allowing a simultaneous view of both.

The horizontal slots receive vertically polarized radiation and the vertical slots receive horizontally polarized radiation.

Consider two nearby antennas that are orthogonal to each other. They look at nearly the same spot in the sky and receive the same sky polarization.

If the horizontal antenna detects maximum power and the vertical antenna detects zero power, the incident sky polarization is vertical.

If the horizontal antenna detects zero power and the vertical antenna detects maximum power, the incident sky polarization is horizontal.

If both antennas detect the same power, the incident sky polarization is at 45 degree angle (or 135 degree angle). Or the sky radiation is unpolarized (randomly polarized.)

When the two antennas record different amounts of nonzero power and the fraction of polarized intensity is known, the sky polarization angle can be calculated.

Any ambiguity in the angle is resolved by adding other information and coordinating across the entire imaging plane. Note that the antennas divide the total incoming radiation into horizontal and vertical polarization. The intrinsic polarization BICEP2 is searching for is over and above these polarizations.

236

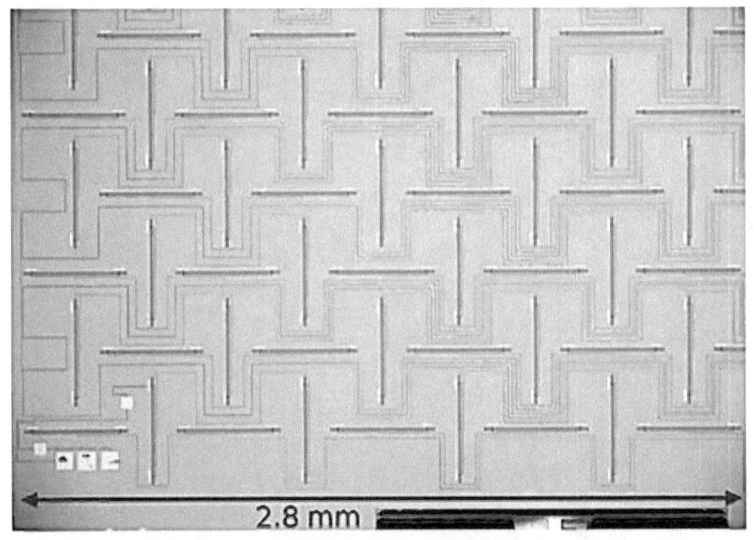

(a) A portion of the BICEP2 focal plane slot antenna array

(b) A portion of the above array within a circle of radius equal to λ/3 about the center X (λ= the wavelength = 2 mm)

Figure IV-22: The spacing of the slot antennas in the imaging plane array.

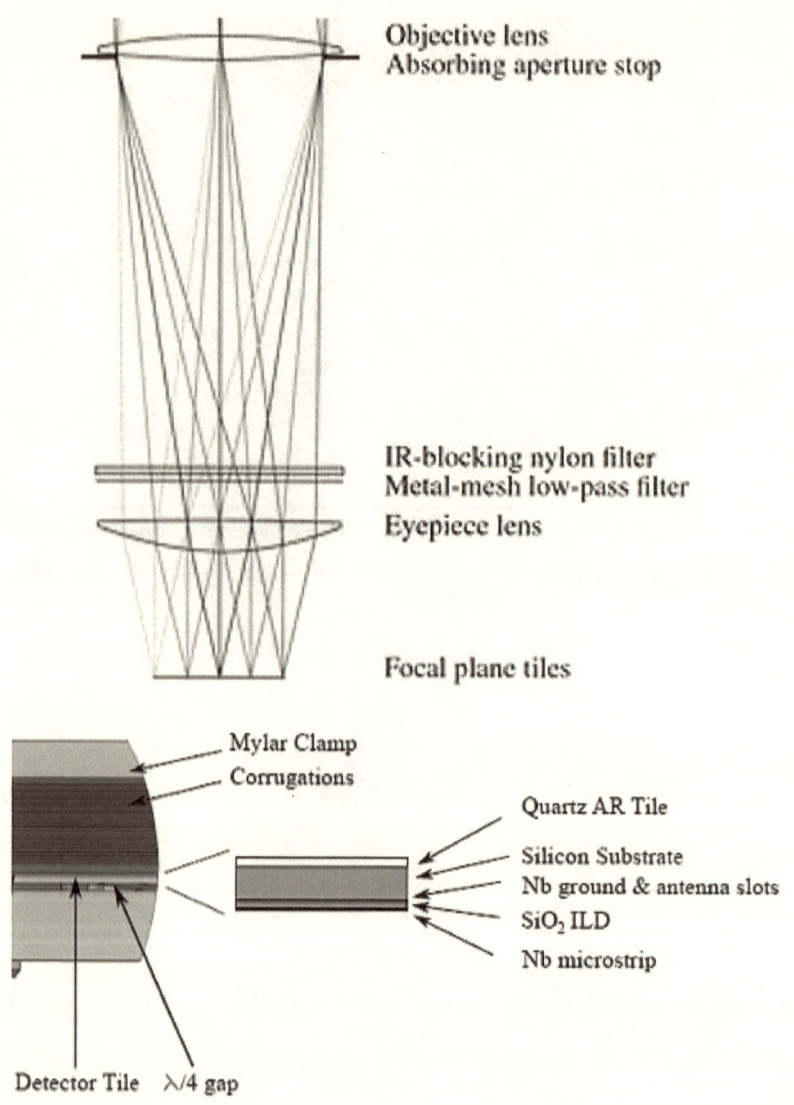

Objective lens
Absorbing aperture stop

IR-blocking nylon filter
Metal-mesh low-pass filter

Eyepiece lens

Focal plane tiles

Mylar Clamp
Corrugations

Quartz AR Tile

Silicon Substrate
Nb ground & antenna slots
SiO$_2$ ILD
Nb microstrip

Detector Tile λ/4 gap

Figure IV-23: *Top* – BICEP2 telescope ray optics showing how families of rays arriving at different angles are brought to different points in the focal (imaging) plane. *Bottom* – The framing around the imaging plane, and an expanded view of the imaging plane layers. (*Image source: BICEP2 Collaboration.*)

This is the very basic principle of BICEP2 polarimetry we need to know for our discussion. Of course the practice is far more complicated. The main points for our purpose are:

1. Each location a in the focal plane corresponds to a location A in the sky. If a moves, A moves.

2. The antennas must be identical in their electromagnetic properties. What this means for our specific purpose is that for the same amount of co-polarized power incident on an antenna, each antenna must report exactly the same amount of energy. When I say exactly, I mean there is very little tolerance, probably only a small fraction of 1%.

Note that BICEP2 telescope can be rotated about its axis. The imaging plane is rigidly fixed to the body of the telescope so that it rotates with the telescope.

Refer now to Figure IV-23. Above the metal plane containing the slot antennas is a seemingly fairly thick (not negligible compared to a wavelength) layer of dielectric (silicon) having a dielectric constant $\epsilon \sim 11.8$. It is actually the substrate for the metal plane. (In conventional usage the substrate would be at the bottom and the bare antenna metal plane on top.)

This is a problem. The substrate – if it was to face the incoming radiation - should have been a thin layer of low dielectric constant (~1) material. The silicon layer strongly bends the incoming rays that arrive from off-vertical directions in a way that has not been figured into the telescope optics. Note that the refractive index of this dielectric layer is high ($n = \sqrt{\epsilon} = 3.44$). This bending has the effect of diverting rays off their destination slot. So what would be the point of spending millions of dollars and spinning all this rigmarole to collect the faintest of faint radiation, only to waste some of it at the focal plane?

There is another problem. There is a quarter-wave anti-reflection (AR) quartz layer on top of this substrate for reducing reflection at the air/silicon interface. Such a layer affects off-vertical rays in a way that has not been included in the analysis.

There is also a quarter-wave layer at the bottom of the microstrip layer. If these quarter-wave layers have been placed in the near field of the antennas, then this is an inappropriate procedure. The near field is a non-propagation region where the wavelength has no clear meaning.

IV-6.5 Antenna fault

A discussion of antenna fault should best begin by reminding the reader what was being asked of the antenna. The design goal was to achieve 1 part in 30 million sensitivity in measuring the power incident on the telescope, and a determination of the polarization vector accurate enough to show the coherent polarization swirls in the sky.

The first antenna fault is the size of the slots. The length of a slot appears to be about 0.4 mm, judging from the pictures. Since the wavelength is 2 mm, we have an antenna of size $\lambda/5$. Technically, this not quite an *electrically small antenna* (for which the size is $\leq \lambda/10$.) And yet its performance has begun to degrade towards the properties of such an antenna and away from those of full size antennas (e.g., a length $\geq \sim \lambda/2$.) This antenna therefore has degraded polarization characteristics.

The above discussion is next complicated by the presence of the high dielectric substrate. The effective wavelength λ_{eff} in the substrate is smaller than the free space wavelength λ ($\lambda_{eff} = \lambda/\sqrt{\epsilon}$). One might thus think that one is dealing with an electrically full size antenna. But this is not the case, and a discussion of this becomes rapidly complex. Suffice it to say that such antennas have not been widely studied, and should not have been included in a scientifically characterizable system.

Secondly, the design of a slot antenna requires that the microstrip transmission lines (whether used for radiofrequency transmission or simply detected voltage) on the circuit board stay clear of the physical slot to some distance around it – as shown in the case of the vertical antennas. However, this principle was not followed for the horizontal antennas.

As a result, the properties of the horizontal and the vertical antennas are different (Figure IV-24). They will report different amounts of power when the same amount of co-polarized power is incident on them. As I have explained, such a difference translates to a false polarization angle ascribed to the incoming sky wave.

Thus an intrinsic instrumental polarization is introduced at all circuit board locations a, and are falsely ascribed to all corresponding sky points A being observed.

Furthermore, the said difference in the power varies across the plane because of the way the circuit has been designed.

So the instrumental polarization has an entire polarization map (mosaic) of itself. If there is any native polarization in the sky, the map BICEP2 obtains is some kind of convolution of the instrumental polarization map and the sky polarization map.

If the focal plane (i.e., the telescope) is rotated about the telescope axis, the location of a in the telescope changes with respect to the sky. And so the location A in the sky changes. Thus the *instrumental* map of the sky rotates with the telescope while the sky polarization map remains fixed in the sky.

Therefore, when the telescope is rotated, the sky image reported by BICEP2 structurally rotates. This would be the most crucial test to conduct with the BICEP2 telescope before any observations are made.

241

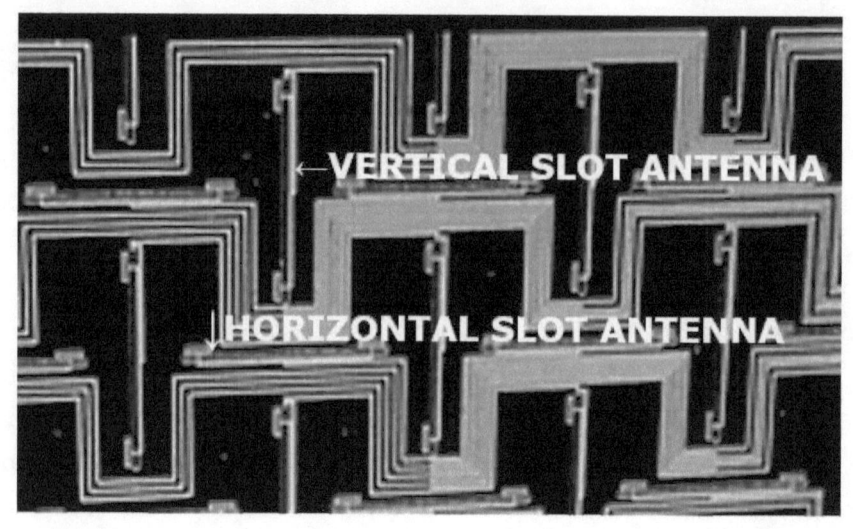

VERTICAL SLOT ANTENNA

HORIZONTAL SLOT ANTENNA

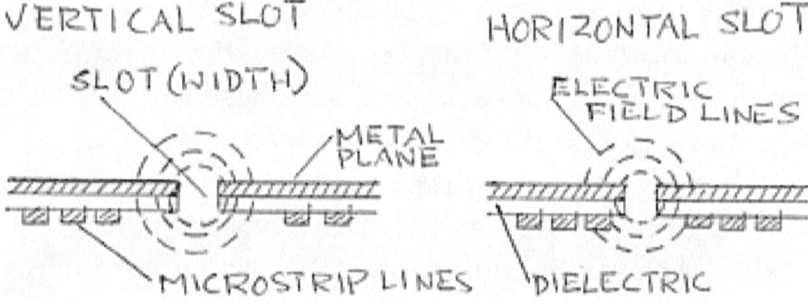

VERTICAL SLOT

SLOT (WIDTH)

METAL PLANE

MICROSTRIP LINES

OK FOR SOME PURPOSES

HORIZONTAL SLOT

ELECTRIC FIELD LINES

DIELECTRIC

NO NO!

Figure IV-24: *Top* – BICEP2 imaging plane slot antennas and microstrip lines to the antennas. *Bottom* – Vertical sections of the plane showing construction of the slot antennas and the microstrip line layout next to the antennas. The silicon dioxide interlayer dielectric serves to insulate the metal plane from the microstrip lines. In antenna design practice it is forbidden for the stripline to encroach this close into the active region of the antennas.

This was never done, indicating clearly that the designers were not even aware of these fundamental scientific issues.

Remember that BICEP2 was looking for the smallest of small signals. There needed to be not even an appearance that the horizontal and the vertical antennas are different. Instead, we have this clear design violation.

To summarize the discussion thus far:

1. Given the sensitivity required of the instrument, the horizontal and the vertical antennas are not the same in terms of their overall Gain and radiation patterns.

2. Because the antennas are much smaller than a wavelength and as such not well characterized, they are unsuitable as components in a precision scientific measurement instrument.

3. The incident polarization angle will be rotated by the instrument because of the dissimilarity between the adjacent horizontal and vertical antennas.

4. Because of the layout of the microstrip circuitry, the focal plane has a clear directionality, the x and the y axis being not interchangeable.

5. The focal plane properties become convoluted with any actual skymap the telescope is observing, to provide an artifactual skymap.

In many popular discussions of the BICEP2 telescope, I have seen comments confusing antennas and detectors. It is said that the telescope is so phenomenal because it has so many detectors in the imaging plane, each detector being of high quality. The quality of imaging is determined by the antennas. They determine the optics. If this is not done right, it is immaterial what the detector quality is and how many detectors there are.

This confusion is also evident in invoking digital camera

analogies (pixels etc.) to the BICEP2 class telescopes. The researchers frequently mix up optical and microwave concepts. Devices that are at odds have been slapped together willy-nilly.

Statements were made that the metal plane in which the antenna slots are cut shields the microstrip circuitry from the incoming electromagnetic wave. As I have shown in Figure IV-24, this simply cannot be the case.

IV-6.6 Polarization fault

In Section III-1.8 I discussed polarization pattern and axial ratio of a linearly polarized antenna. These determine how well polarized an antenna is. The higher the axial ratio, the greater the degree of linear polarization.

For the BICEP2 antenna elements, these test data for the telescope-mounted individual slot antenna have not been presented. But one can estimate that for this application the axial ratio has to be better than ~ 40 dB. The actual axial ratio of a BICEP2 antenna element is probably around 20 dB, and certainly no better than ~ 30 dB. With this axial ratio it would be impossible to detect and map the B-mode polarization swirls in CMB that have been reported.

IV-6.7 Array fault

Let us now turn to the antenna array, a portion of which is shown in Figure IV-22(a). The progress of the BICEP2 class telescopes has been driven by packing more and more antenna elements into the same focal plane area, reportedly to provide higher and higher resolution and faster and faster imaging of the sky.

In Figure IV-22(b) is shown one slot antenna of the array (marked by an x) with a circle drawn around it. The radius of this circle is $\lambda/3$ (λ= the wavelength, in this case 2 mm). Now we can discuss the situation from the point of view of basic physics and then, almost equivalently, from the point of view of antenna theory.

The physics view: One of the fundamental limitations of Electromagnetic Theory is that electromagnetic waves cannot resolve (discern, discriminate) structures that are closer than ~ $\lambda/3$.

What this means in the present context is that to the incoming wave, the central antenna and any other antenna within the circle in Figure IV-22(b) are indistinguishable. The wave sees the whole circle as one blurred area. Therefore the theory that each slot in the imaging plane (or a group of slots) distinctly images a single area of the sky is emphatically wrong.

Packing the slots closer and closer does not lead to greater and greater definition of the sky after a certain separation distance between the neighboring antennas has been reached. Yet, from many statements in evidence, this is what the BICEP2 Collaboration thinks to be the case.

The antenna theory view: From the point of antenna theory the circle represents the near field of the central antenna. Therefore the antennas within the circle are not independent of one another. When they needed each to be strictly isolated, there is instead cross-talk among them by virtue of bad design. Hence they are neither suitable for imaging nor for polarization measurement.

The principle of BICEP2 operation requires first and foremost that each antenna be independent of its neighboring antennas, that it have a high isolation. In actuality, the antennas in the entire imaging plane are interconnected through interference between contiguous regions such as shown in Figure IV-22(b).

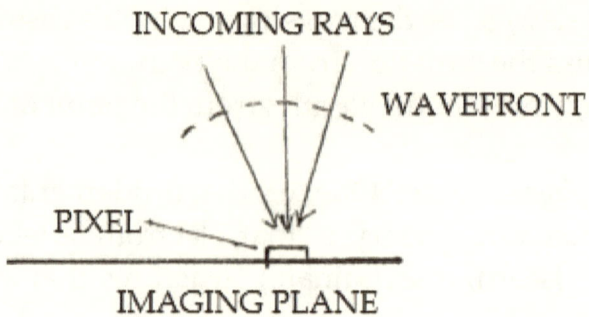

INCOMING RAYS

WAVEFRONT

PIXEL

IMAGING PLANE

Figure V-25: BICEP2 used the planar wavefront phased array concept whereas the actual wavefront is spherical (cf. Figure IV-23).

Furthermore, the planar antenna array theory used in BICEP2 design is inapplicable. The array theory used assumes a plane wave (equal phase front) incident on the array at some angle to the forward direction. This is not the case with BICEP2, where the incoming wave has a spherical-like wavefront (Figure V-25).

I have come across a paper which spells out how the antenna spacing was arrived at (see Figure IV-26 which is self-explanatory.) It mistakenly uses a value of 11.8 for ϵ_r when the correct value to be used is 1.0. So the mystery of how principles of physics and antenna engineering were violated by a large factor is solved.

A physicist here can use common-sense to see how odd this calculation is. If the silicon dielectric layer over the antenna slots (refer back to Figure IV-23) restricts the antenna spacing so drastically, then Equation (1) of Figure IV-26 should include the thickness of this layer as a parameter. Or this thickness must be much greater than the wavelength. So even on a cursory look this calculation is wrong. And yet, today in the spring of 2015, this is the operating principle of a number of telescopes, out there in the field trying to make grand discoveries.

3.1. *Antenna Design*

The antenna slots in each detector must be spaced to Nyquist sample the focal plane surface to avoid grating lobes that would rapidly change the impedance with frequency (Kuo et al. 2008). The antenna pattern of each axis of an array is calculated from the N elements per linear dimension spaced at distance s as follows:

$$A(\theta)=\sum_{m=-(N-1)/2}^{(N-1)/2} e^{-j2\pi\frac{m\,s\,\sqrt{\epsilon_r}}{\lambda_o}\sin(\theta)}$$

$$=\frac{\sin(N\pi s\sqrt{\epsilon_r}\sin\theta/\lambda_o)}{\sin(\pi s\sqrt{\epsilon_r}\sin\theta/\lambda_o)}, \quad (1)$$

where λ_o is the free-space wavelength, ϵ_r the relative permittivity of the surrounding medium, and the sum is across sub-antennas indexed by m. In addition to the strong peak in the normal direction ($\theta = 0$), there are grating lobe peaks when $\sqrt{\epsilon_r}s\sin(\theta)/\lambda_o$ is a positive integer. To avoid these lobes, the slot spacing must be

$$s \leq \frac{\lambda_{o,min}}{\sqrt{\epsilon_r}}\left(1-\frac{1}{N}\right), \quad (2)$$

where $\lambda_{o,min}$ is the minimum wavelength of operation and the term in parentheses accounts for the finite width of the grating-lobe peaks. For the 150 GHz detectors fabricated on silicon ($\epsilon_r = 11.8$) with an upper band edge of 180 GHz ($\lambda_{o,min} = 1.7$ mm), the spacing must satisfy $s \leq 460\ \mu$m. To

http://arxiv.org/pdf/1502.00596.pdf

Figure IV-26: BICEP2 team's calculation showing how the BICEP2 antenna spacing was arrived at.

IV-6.8 BICEP2 observation technique

It is clear that the BICEP2 team was not even aware of the issues I have discussed above. They have in fact bandied around images of the focal plane all over the place with great parental pride. These images contained clear visual signal of what was wrong.

However, it seems that they concluded that some type of attention needed to be given to the angular position of the telescope about its axis. This is how BICEP2 team leader James Bock of California Institute of Technology described this:

To make accurate measurements over a wide area, the challenge is to control false signals. ... Finally, to remove from the system any effects that might arise from having a preferred direction, we spin our telescope around its axis every day.

So it seems that some type of angle-averaging or angle randomization with regard to some unknown suspected directionality in the telescope imaging plane underlies the BICEP2 sky maps unveiled.

This is most curious in a state-of-the-art, pioneering experiment attempting to make the grandest of discoveries by pushing the limits of measurability.

If there is a suspected directionality, would one not want to examine and pinpoint this? Especially when it takes no more effort than making sky images with the telescope fixed at 0, 45 and 90 degrees position (for example.)

What does it mean exactly to average maps if they were structurally rotating?

The BICEP2 team had the burden to produce those angle-specific skymaps before publishing their discovery. This crucial burden on the BICEP2 experimenters cannot be avoided with statements like *We compared BICEP2 with other telescopes and everything is fine; We are very careful scientists; We have worked very hard for years;* and *B-mode polarization is on the sky.*

Where the instrument on the ground is clearly at fault physics-wise and engineering-wise, to look for evidence in the quality of the skymaps themselves that everything is fine with the instrument on the ground is a non sequitur, to say the least.

IV-6.9 The proof offered: Bicep1 vs Bicep2

Let us nevertheless consider the self-explanatory Figure IV-27. Here we see that for the E-mode polarization, there is some correspondence between BICEP1 and BICEP2 (Not so for B-mode). Specifically, the directionality of the features in the map seems to be the same.

Note that BICEP1 did not have the grid type focal plane. It had the tried and true dual polarized horns in its focal plane. But this is not to say that everything else was fine with BICEP1. This is not the subject of the present investigation. To summarize:

We end up with a stark contradiction between basic physics principles and basic antenna design principles in the textbooks on one hand, and the phenomenally precision astronomical observations reported by the BICEP2 team on the other:

SCIENTIFIC DESIGN ANALYSIS: The telescope is not capable of measuring polarization features in the CMB radiation because of its small aperture, and faulty design of the antenna elements. The telescope's imaging plane has a pronounced directionality, and adds an instrumental polarization component across the skymap. Tighter packing of antennas cannot lead to higher resolution beyond a certain point – which was crossed by a great extent.

COMPARISON BETWEEN BICEP1 AND BICEP2
How the bicep2 imaging plane modifies sky image

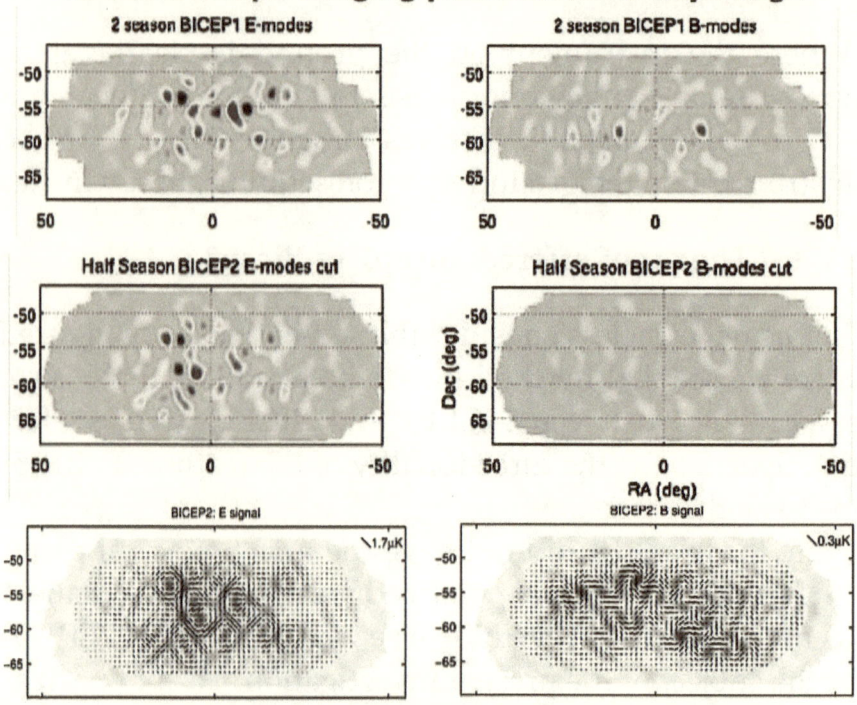

WHAT THE MAPS ARE POSITED TO BE BY THE BICEP2 TEAM:

TOP PANEL: Sky as fully imaged by bicep1.
MIDDLE PANEL: Sky as imaged by bicep2 for half season.
BOTTOM PANEL: Sky as fully imaged by bicep2.

WHAT THE MAPS ACTUALLY ARE:

TOP PANEL: Sky as fully imaged by bicep1.
MIDDLE PANEL: Sky as modified by bicep2 imaging plane for half season
BOTTOM PANEL: Sky as fully modified by bicep2 imaging plane.

Figure IV-27: Comparison of BICEP1 (with dual polarized horn antenna arrays in the imaging plane) and BICEP2 (with the experimental slot antenna array in the imaging plane) skymaps.

OBSERVATION REPORTED: The telescope is reporting with extreme confidence crisp, tightly woven high resolution polarization skymaps that do not depend on the angular orientation of the telescope about its axis.

IV-6.10 Image presentation

When the BICEP2 sky images of polarization in the Cosmic Microwave Background radiation were unveiled, they were plotted on a Right Ascension-Declination diagram with different scales for the different axes. On this diagram, the full moon would look like a vertical bar. In Figure IV-28, I have resized one such diagram to where the full moon would look like a circular disc.

And what do we see here? There was a preference of the gravitational waves produced during the inflation era 14 billion years ago to favor the Earthly coordinate system. A coincidence? Let us explore further.

We can take a red pencil and mark as many right angles as can be found. All of them are perfectly aligned with RA-Dec coordinate system. Another coincidence? No. These are clear and expected results of the BICEP2 instrumental botch-up.

IV-6.11 The botch-up on-the-sky

Since the imaging plane is manifestly faulty in so many basic ways, there is no need to look to the sky images for signs of the botch-up. However, when there emerged comparison skymaps between BICEP2 and the Keck Array telescopes for the very time, something very strange was immediately evident (Figure IV-29).

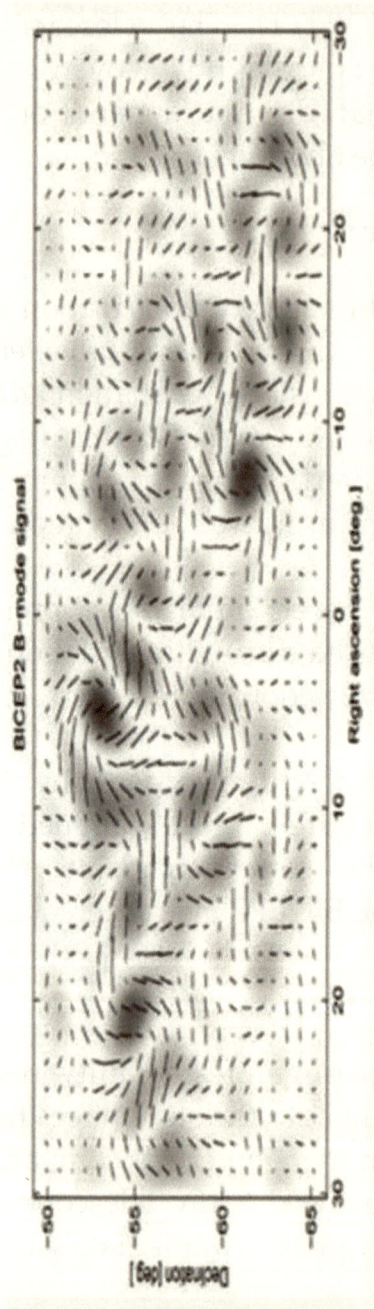

BICEP2 B-MODE SKY SWIRLS PLOTTED ON THE SAME RIGHT ASCENSION AND DECLINATION SCALE. NOTICE THAT THE POLARIZATION SWIRLS HAVE A PRONOUNCED DIRECTIONALITY: THEY HAVE AN ELLIPTICAL TENDENCY WITH THE LONG AXIS ALONG A CONSTANT DECLINATION LINE. THE B-MODE POLARIZATION FAVORS THE ORIENTATION OF OUR LITTLE PLANET!

Figure IV-28

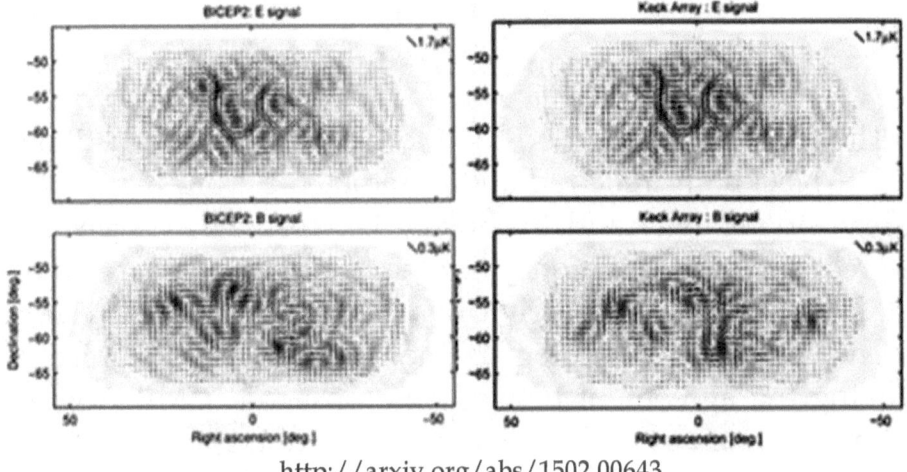

http://arxiv.org/abs/1502.00643

Figure IV-29: Comparison between BICEP2 (left) and Keck Array skymaps of E-mode (top) and B-mode (bottom) signals.

Here the two telescopes had exactly the same imaging technology and were observing exactly the same spot of the sky from exactly the same spot on the Earth at exactly the same frequency. The E-mode skymaps are exactly the same to the minutest detail (examine the original figure), and the B-mode skymaps are altogether different. It is hard to say what is going on without knowing the relative orientation of the imaging planes. But there can in any event be one and only one conclusion: instrumental defect.

The BICEP2 team ushered in a new era of telescope science. They kept presenting assorted polarization maps and averring that the images are "on the sky", "on the sky" … (meaning they are true sky features, and not instrumental artifacts.) When erroneous physics and engineering are plainly evident in the instrument on ground, how one could point to sky images to assert that everything is fine is something I never learned during my stint in radio astronomy or satellite communication.

CHAPTER IV-7
The second CMB deception

The first CMB deception tried to pass off CMB radiation as the 3 K blackbody relic radiation predicted by Big Bang. With the exposing of that fraud, the linkage between the observed and the predicted relic radiation was gone. There was no basis to say that CMB represents the early Universe or the Baby Universe.

The Big Bang cosmologists then provided a reprieve by invoking the agreement between the observed CMB Angular Power Spectrum and that modeled from the inflation theory (Figure IV-30). This agreement is now said to be the proof that CMB represents the Baby Universe.

This scam is rather subtle. The theoretical curve was set by events in (the parameters of) the inflation era, which events would be imprinted in the theorized relic radiation then emitted. At some point after the inflation era, this relic radiation is theorized to have assumed the blackbody form – still carrying the imprint. The theory in Figure IV-30 starts from the assumption of the blackbody (thermalized radiation.) So the theory line in the diagram can be compared with the observed CMB *if and only if* the latter has the blackbody form.

But the blackbody form of CMB radiation has been repeatedly falsified by satellite observations. The data points in the diagram are not from a blackbody. So this theory-experiment comparison is a comparison between apples and oranges. The impressive agreement is the scam. The true CMB spectrum is accurately known from the WMAP and Planck satellites, and it can be inferred that this spectrum could not be more different from a blackbody. However, as mentioned before, this spectrum has been embargoed so that the triumphal march of Big Bang can continue.

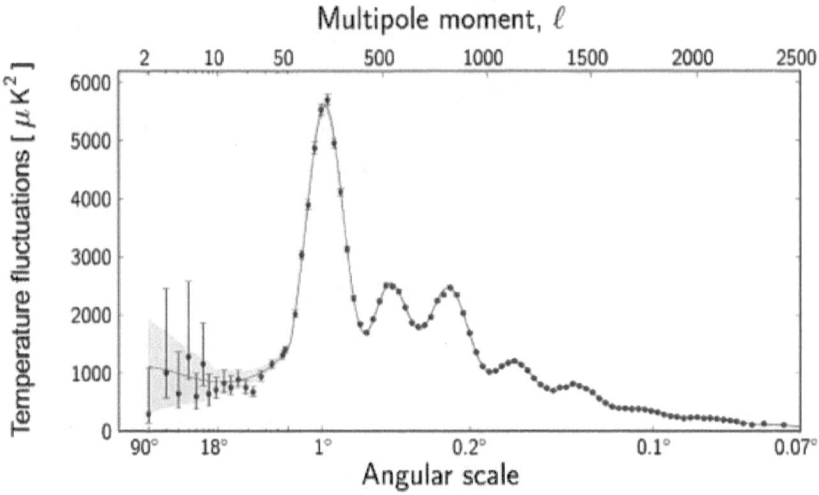

Figure IV-30: The CMB Angular Power Spectrum. The angular scale
in the abscissa shows angles in the sky that are harmonics
(represented by the Multipole moment l) of the 360 degree sky.
Thus an angle of 90 degrees corresponds to two full cycles in the sky
($l = 2$). The ordinate represents the temperature fluctuations over
and above the CMB blackbody temperature of 2.7 K, on the given
angular size. The dots are data points derived from the observed
CMB skymap. The continuous line is the theoretical modeling from
the inflation theory.

But how did this impressive agreement come about? When
you have in your theory spherical harmonics, Fourier series etc.
and otherwise enormous leeway, the above fit to the inflation
theory is not at all surprising. A good practicing physicist
should find nothing compelling about the agreement. But others
may find this most compelling.

One should note also that the observational data points in
this diagram are not the measured point-by-point intensity in
the sky – but have gone through some *ad hoc* processing to
falsely convert the intensities into temperatures. In other words,

the theory (based on blackbody) and the data (from a non-blackbody) are being made to meet each other halfway.

When one flies in an airplane in daylight hours, one will often see below a cloud cover with a patchy structure much like the anisotropic CMB sky. It is quite possible that inflation theory can be made to "predict" that as well with back-calculated theory parameters.

Likewise, there may be generated a whole series of fractal anisotropy skymaps which will also corroborate inflation theory.

Indeed, writing in the journal Nature on 3 June 2014, Princeton University cosmologist Paul Steinhardt, a Big Bang insider, explains:

...the inflationary paradigm is so flexible that it is immune to experimental and observational tests. First, inflation is driven by a hypothetical scalar field, the inflation, which has properties that can be adjusted to produce effectively any outcome...

What this says in concrete terms is that the fit in Figure IV-30 is meaningless, the enormous volume of scientific literature to the contrary notwithstanding.

> With four parameters I can fit an elephant, and with five I can make him wiggle his trunk.
>
> John von Neumann

CHAPTER IV-8
CMB: A fresh look

IV-8.1 Why a fresh look?

In the Big Bang frequency and temperature regime, 200 GHz and 2.7 K, say, $h\nu/kT \sim 3.6$. So let us for convenience say $\exp(h\nu/kT) \gg 1$ (see Table III-1). In this approximation

$$[\Delta B_\nu/B_\nu] \approx (h\nu/kT) [\Delta T/T].$$

This is a simplified representation of the method by which an observed blackbody intensity variation ΔB_ν can be converted to a temperature variation ΔT to obtain the Big Bang skymap.

In the actual sky there is no blackbody, and left hand side of the above relation is taken from the intensity measurements ΔI_ν. So the use of this equation is patently wrong.

There is also no temperature T in the sky but it is force-fitted to 2.7 K anyway. This shows how precision satellite measurement data obtained at great cost are put through a two-step corruption and placed in the service of Big Bang.

So we need to consider CMB afresh: What is CMB in itself? What is its nature?

IV-8.2 CMB: What is it in itself?

Cosmic Microwave Background radiation is relic radiation only within the framework of Big Bang theory. According to this theory, the CMB we see today arose 14 billion years ago, and is coming to us from distances of tens of billions of light years. It is the light of the early Universe.

However, decades of application of mass psychology techniques by the Big Bang cosmologists has inured us to the view that CMB ≡ relic blackbody as a universal truth. It is only when we deprogram ourselves from this jaundiced condition that we can undertake a proper scientific discussion of CMB strictly as an observed phenomenon, and strictly as a matter of physics.

What is this CMB in itself? It is an all-pervasive radiation coming to us from all directions at all times, spanning the frequency range (as far as we know today) from the red edge of the optical region through the infrared region to the centimeter wave microwave region. Whether the CMB frequency range extends further in either direction is not known at present.

The true CMB spectrum in the Big Bang blackbody frequency window is known to the Planck Collaboration with great accuracy, and has not been released. Likewise the sky distribution of CMB observed total intensity (as distinct from "temperaturized" intensity) is known to the Planck Collaboration and has not been released. The reason for this embargo is almost certainly that this information would falsify Big Bang cosmology from the beginning to the end. Because of this stratagem our objective discussion of CMB is necessarily more speculative today than it has to be.

This radiation is observed to be isotropic, meaning that it surrounds us in the celestial sphere like overcast sky. However, a very small fraction of this intensity – perhaps 10-100 millionths – has a patchy structure like cumulous clouds. This patchy structure is fixed in the celestial sphere.

There are reports of observation of polarization of radiation in this very faint patchy structure, but these measurements are highly questionable at the present time.

IV-8.3 CMB: What is its nature?

The view promulgated by the Big Bang cosmologists that the discovery of CMB by Penzias and Wilson in 1964 was a great surprise is most disingenuous. There is absolutely nothing surprising that there should be some radiation at some frequency in the sky. There are all kinds of mechanisms of emission, absorption and reradiation in the Universe. What would be surprising is if someone discovered a frequency window in the sky which there is no radiation. In this sense, CMB is a completely expected finding. We will consider three known properties of CMB.

Frequency span

There are frequency conversion mechanisms in the Universe that can spread out the radiation emitted by localized sources over a broader frequency range. A star absorbs energy from all kinds of sources incident on it at all kinds of frequencies, but emits a fixed spectrum. Thus it helps homogenize the spectrum of radiation due to various emission mechanisms. The same is true to some extent for interstellar dust and regions of ionized gases. These processes both upconvert and downconvert radiation.

Isotropy

One of the strongest points Big Bang cosmologists offer to support the assumption that CMB is relic radiation is that CMB is isotropic. There are no other mechanisms by which such isotropy could have arisen, they say.

But the radiation emitted by localized sources are scattered by the interstellar dust grains (and also by ionized gases) in a range of directions. It is in fact the finding of the Big Bang satellite Planck that dust exists in every direction in the sky. It follows then that CMB radiation is coming from every direction in the sky. There is no need to invoke a relic source argument to explain the isotropy.

Elongated interstellar dust can polarize the unpolarized radiation from localized sources through the same process of scattering. Electrically charged interstellar dust emit radiation by virtue of their spin motion, and this radiation can be partly polarized.

The above mechanisms are ones that we know are occurring. They have to be reckoned with before slipping in the Big Bang explanation of CMB.

The patchy structure is a minuscule effect, but is worth studying. How far is CMB, how old is CMB – these are questions that are completely open for discussion.

In Appendix D, I propose a mechanism which may be causing the radiation from localized sources in the universe to continually shift towards lower frequencies. This may do away with the stretching-of-space (expanding Universe) explanation of the observed cosmological redshift.

IV-8.3 Conclusions

To summarize, there are known physics mechanisms that must necessarily be occurring and that result in the electromagnetic radiation filling the Universe to continually

- shift upward or downward in frequency;
- have a component that is isotropic;
- assume some degree and state of polarization.

Just as the sky is blue because of scattering of sunlight by atmospheric molecules, so perhaps is the celestial sphere CMB because of scattering and re-radiation of starlight by interstellar dust.

THE BOOK OF VERDICTS

The most practical function that an investigation of some manmade disaster or scourge can serve is the assignment of blame. And when public funds and public interest are involved, such assignment must be made in direct language.
The breadth and depth of easy dishonesty among the practitioners of Big Bang cosmology may be a finding unprecedented in a community that wears the noble gown of trustworthiness and respectability.
But it is far worse for their support groups.
The very people whose responsibility it was to deal with the shamming and scamming and defrauding by the Big Bang cosmologists are the very ones that aided, abetted and covered up.
There is a Bengali proverb about sleaziness:
Lewd dancing going on behind the veil of modesty.
Here we have an entire phantasmagoric orgy being played out on the world stage – among the noblest of noble intellectuals. Taxpayers are footing the bill.

Enlightened people seldom or never possess a sense of responsibility.

George Orwell

CHAPTER V-1
Penzias and Wilson

Arno Allan Penzias and Robert Woodrow Wilson were accidental discoverers. They never claimed to be anything else. And they truly were the discoverers of the Cosmic Microwave Background radiation.

But they were manipulated by Princeton's Robert Henry Dicke, Philip James Peebles and others into misinterpreting their data to serve the cause of Big Bang cosmology. They were posited as the discoverers of the Big Bang relic blackbody radiation. Penzias and Wilson should have known better than to allow themselves to be manipulated thusly. In this sense they are at fault.

They seemed genuinely surprised and humbled by the 1978 Nobel Prize. They both acted with dignity and reserve following their achieving the Nobel Laureate status. They do not seem to be in evidence pushing Big Bang cosmology.

Robert Dicke was an expert in antenna and microwave theory and techniques. He had in fact invented the radiometer known today by his name. He had long done experiments in this area. He worked with experimental groups in this area at the MIT RadLab. He and his experimental colleagues at Princeton had to know that they were force-fitting the Penzias-Wilson observation into the Big Bang theory. They probably felt that somehow getting this Big Bang confirmation into print in a peer-reviewed journal would legitimize it and place it beyond question. It did. The role of *Astrophysical Journal* in publishing the twin-papers is to be noted here – as also with other bogus Big Bang discoveries. With the discovery thus cordoned off within the perimeter of astrophysics, there was never any occasion for any appropriate engineering experts to come to scrutinize it. Which industry engineer has the reason and the time and the inclination to go digging into cosmological discoveries consummated by high academics? The discovery was saved harmless from the prying eyes of true experts.

There was talk in the scientific circle of Dicke getting a share of the Nobel Prize. After all, three people can share the Physics Nobel Prize in any given year. But for some reason that did not happen.

At the time of closing this book, Arno Penzias seemed to be in quiet, well-earned retirement. Robert Wilson held a position at the Harvard Smithsonian Center for Astrophysics and was very much in evidence in the Big Bang scene. He participated in scientific meetings, and spoke of his discovery. He gave interviews. He was very visible at the Press Conference where the BICEP2 discovery was announced on 17 March 2014. However, in all these it seemed that he was generally enjoying himself and not pushing or selling anything.

History might adopt a revised view that Penzias and Wilson are discoverers of CMB – an important observational discovery. In this way the two men may be held blameless as to Big Bang while at the same time given a continued place in the pantheon of the discoverers. Whether or not these scientists want to make public statements disowning Big Bang is of course their choice. It is after all their history and their legacy.

CHAPTER V-2
John Cromwell Mather

V-2.1 Deceiving science and society

John Cromwell Mather is a public figure and a scientist on public payroll. He has been on public payroll all through the period under discussion here. In these capacities he has deceived science and society at three distinct levels:

Expensive scientific bungling, ineptness;
Scientific fraud to cover up above and then glorify the result thus obtained;
Public promotional campaign to cover up the fraud when exposed.

The first two levels I have covered in Chapter IV-2. The third level concerns the unprecedented spectacle of public feting and worshipping of Mather that went on since I exposed him to be a science fraud. These I will cover under the various chapters of Book V.

V-2.2 The public denials

Following my exposing of the fraud in April 2007, John Mather had a number of options on how to admit fault, accept responsibility, and move on. Given the extremely powerful circles of friends he has in the academic, government and legislative spheres – all well-connected to the media – he would have been able to minimize the damage to himself. His clever friends might even have been able to spin it as a positive thing. But instead Mather has continued to this day to maintain that his discovery is completely intact.

In December 2008 in his Commencement Speech to the

graduating class of the University of Maryland he said:

... We are going to be found out if we try to cheat...if I publish something that is wrong, my friends are going to find out. If I find out it's wrong before they do, I'll publish a correction.... For a scientist, there is only one easy path, the path of scrupulous honesty. Honesty is relaxing. There is nothing to hide and nothing to fear.

In 2009 Mather inaugurated the Nobel Youtube program named *Ask a Nobel Laureate*. There the following question-and-answer session took place:

High school student: I was wondering, have you had many people trying to contradict your theory that won you the Nobel Prize?

John Mather: Hi -----! Actually it's not my theory – but my measurements that got a prize for us. So... no, I don't think anyone has tried seriously to contradict our measurements. There are a few people who don't think we did it right but there are only a few such people.

In the summer of 2012, Mather participated in an event of the World Science Festival along with a supporting cast of a number of prominent Big Bang scientists on the stage. There he told the audience that even though "thousands" of (critical) papers were published on the George Smoot skymap, not a single paper was published on Mather's blackbody. The clear implication was that the blackbody stands unquestioned by the scientific establishment.

During this entire time Mather remained in public view – making frequent appearances in the lecture circuit, the festival circuit, and so on. He probably was more visible than any other Nobel Laureates in physics. In all such occasions he exuded an aura of absolute self-assuredness and total confidence.

One of the principal themes of these frequent public appearances has been that he is out and about to inspire children and young people with his own example.

Anyone observing him in these arenas over these many years would not even remotely suspect that anything was amiss.

V-2.3 The finding of fraud

There exist today a few instances, right there within the physics establishment, of reporting wrong discoveries that were either investigated by the establishment or where the discoverer retracted.

In the cases of Jan Hendrik Schön and Victor Ninov, the establishment made specific determinations that science fraud was committed. In the instance of P. Buford Price and his discovery of magnetic monopole, he retracted before any criticisms arose. There are also examples of retractions following airing of criticism. So we have a complete established framework within the physicsdom for discussing the discovery of the 2.7 K relic blackbody.

We can say: When an incorrect discovery is reported and the discoverer retracts before anyone spots any errors, it is an honorable act. When criticisms arise and the discoverer retracts, no harm no foul. When an unassailable allegation arises and the discoverer does not retract, there are two possibilities: Ineptness combined with intransigence; and science fraud.

V-2.3 Ineptness defense is off the table

John Mather's own student record foretells, and his own public statements today emphatically aver that he is a supremely competent scientist. His scientific helmsmanship of the next generation space telescope also speaks to this issue. His documented writings also show that he knew what would undo his discovery. And exactly that happened. He noted this problem and dispatched it by hanging on it a ridiculous explanation which he had to know was bogus.

More than seven years of stonewalling and cover up now make it much too late for Mather to 'plead it out' at ineptness. That window has closed.

I can conceive a scenario where Mather's protectors will try to make a case that in his mind Mather truly (honestly) believed all along that he had a discovery. This scenario would let Mather off the hook for the grandest science fraud, and reduce it a simple human failing. Such a scenario might even have been believable and acceptable in the early stage. Now this window too has closed.

V-2.4 Deflecting blame is not an option

So if the ineptness defense is out, what is the only other escape hatch that remains? Point finger to others.

In November of 2013 Mather gave a lecture within the auspices of NASA Goddard Space Flight Center. A video of this lecture, titled *John Mather Maniac Lecture*, was produced. The video was then drawn attention to in a NASA official web page. In the audience that day were present two of Mather's original FIRAS collaborators, Dale Fixsen and Rich Isaacman. It is not known if they came spontaneously or at the invitation of Mather. But Mather certainly planned on their being present.

At about 39 minutes into the video Mather shows his blackbody diagram. About the standing-ovation blackbody he had presented in 1990 he says (as nearly as I can reproduce):

I see Rich Isaacman is here also. He was the software team lead that helped us to figure this out right away.

About his final report of the 50-ppm blackbody, he says:

I see Dale Fixsen here in the audience – he was the person who went from what was a likely blackbody to a blackbody within 50 parts per million. So, a special thanks to Dale for his ability to do that.

While it is good form to give credit to colleagues, it is the particular language that is important to note. Mather is identifying two COBE-FIRAS team members by name. Their presence in the audience gives the names reality. Mather is assigning to each member a particular report of the blackbody. Then he is saying in well-crafted casual language that these people have the ownership of the two blackbodies, not Mather.

This effort should be totally rejected, not because of its cowardice but because of its illogic. John Mather was all along the scientist in charge. Not only was he the leader of the FIRAS project, but he was the overall scientific manager for the COBE satellite. He was the person the scientific establishment considered the point man. What he is doing amounts to a physics experimenter blaming the technician for a discovery that turns out to be bogus.

It was John Mather's scientific quest that had begun long before he came to NASA. He was the Knight looking for the Holy Grail all his scientific life. How can he, upon now finding it, say that the foot soldiers found it. It should be completely unacceptable to the scientific community for Mather to say that there was a gap in the chain of events between the raw satellite data and the final blackbody spectrum for which he is not responsible. Also, over the years he has repeatedly averred (for example, to *Huffington Post*) that he had measured the blackbody "really, really, really well." That statement avers that he was privy to and was knowledgeable on the entire chain of events mentioned above.

V-2.5 Walking away is not on

In December of 2014 something very strange happened. John Mather seemed to make a U-turn from his lifelong scientific path.

First, recall that Smoot and Mather – as subtitled in Smoot's book *Wrinkles in Time* – were both "Witness to the birth of the universe". Mather measured the temperature of the placenta and Smoot made the sonogram picture of the baby. But a ticket event at the Smithsonian Institution on December 16 was advertised as follows:

John Mather, senior astrophysicist at NASA's Goddard Space Flight Center, who will tell you there is no origin to the universe, only a universe that is continually transforming itself.

If this is a stratagem to simply walk away from all that transpired with and around the COBE satellite, it should be found unacceptable. He must first own up to the fact that his 2.7 K blackbody is bogus. Then he must be judged on this the way Jan Hendrik Schön and Victor Ninov were judged by the physics establishment. These two physicists today are known as two most infamous frauds in the history of physics.

V-2.6 A dark chapter of science

From the beginning to the end, FIRAS is an object lesson on how abject scientific quackery can be shepherded to the grandest of discoveries. The entire concept of the FIRAS instrumentation was erroneous to begin with, and stemmed from incompetence. So it matters little that the subsequent engineering study was a series of Inspector Clouseau-style bungling; or that the instrument was launched into space without any scientific characterization on the ground. It should have come as no surprise that once in orbit, the instrument showed fatal fault; or that it never readily produced that telltale null spectrum which was the guiding principle of his instrumentation and the source of its accuracy.

What is surprising is what this completely useless piece of space junk accomplished. First it gave us humankind's most profound Eureka moment: No sooner was the instrument in orbit turned on and some manipulations were performed in the computer than out popped on the NASA Control Center screen the Holy Grail the instrument was looking for – a picture-perfect blackbody spectrum out there in the sky, coming at us from all the way back in time, from the birth pangs of the prenatal Universe. A little later we were to learn about the standing-ovation blackbody with 1% accuracy and a few years later about the 50 ppm blackbody. And let us not forget that he and his collaborators had found the selfsame blackbody in 1975.

Over a period of some four decades, John Mather repeatedly reported discovering something that his instrument was not capable of discovering and that was never there in the sky to discover.

John Mather and his colleagues were succeeding in publishing these bogus discoveries in peer-reviewed scientific journals. It is good thing that this was so. For without this hard documentation it would be most difficult today to open this Cold Case and to make people believe that their superspacehero is in truth a supersomethingelse, given especially the glare of his Nobel Prize.

Slowly abject scientific quackery had morphed into diabolic deceit. Increasingly strident rhetoric about advanced space-age technology and fantastic measurement accuracy was made to swirl around the deceitful discovery. Every power group that

271

could be mobilized to protect the enterprise was mobilized to protect it. Every accolade that could be bestowed to whitewash over it was bestowed to whitewash over it.

It was a very dark chapter of science.

The conduct of the relevant scientific community and the science media surrounding the John Mather matter was darker still.

There is a horror movie called *The Village* where the inhabitants often use an ominous phrase "The things of which we do not speak." This refers to some ghoulish creatures that reside in the forest next to the village and prey upon the villagers. Applied to the aforementioned community, this phrase would refer to the John Mather blackbody, for everyone simply clammed up on this (except for some blackbody die-hards I will discuss later.) When the WMAP satellite results were reported amid great fanfare, no one asked the first and the most obvious question about how the five-frequency data points fell on the blackbody curve. When the Planck satellite results were released after creating an atmosphere of great anticipation, again no one asked how the nine-frequency data points fell on the curve. There truly was a concerted conspiracy of silence. Scientists and the media conspired to deceive the world. As we shall discuss later, the Government and the Legislature became a party to this. And as if all this support was still not enough, the Vatican pitched in.

It was a very dark chapter of the history of scientific civilization: a collaboration of deception among the highest of high intellectuals – the kind humankind had never known before.

And to think that they were all going to get away with this. The Universe was going to stay forever falsified.

CHAPTER V-3
George Fitzgerald Smoot

When it comes to the COBE satellite measurements, no one can accuse George Smoot of ineptness or incompetence. Smoot's resume gives the lie to any such allegations. When he came to the COBE program, he had under his belt all the hands-on experience one might require or desire of him. He had conducted ground-based, balloon-borne and U-2 spy plane-borne experiments precisely in the areas represented in the COBE program. He had also conducted airborne experiments directed at discovering antiworlds made of antimatter. With his high adventurous exploits he himself describes in his book *Wrinkles in Time*, he could be described as the invincible Indiana Jones of space research.

George Smoot was everything that NASA could have asked for in a COBE satellite researcher. In fact, there was no other experimenter more qualified at that time.

It is precisely this unique strength of Smoot that makes it most puzzling why he reported a false discovery.

George Smoot's DMR experiment was physically, electronically and electromagnetically separate from John Mather's FIRAS experiment on the COBE satellite. But as it concerns the underlying theoretical connection, Smoot's discovery was 100% contingent on Mather's discovery having been consummated. If there were no Mather discovery of the relic blackbody, there would be no Smoot discovery of the Baby Universe, regardless of any CMB skymaps Smoot might have found. The reader should recall that the Baby Universe picture – and not merely the CMB skymap – was considered Smoot's actual discovery. It is for this discovery that he was hailed as a hero and given the Nobel Prize.

At the stage in the COBE program we now speak of, Smoot and Mather, as scientific colleagues, were at loggerheads. There was no love lost between them. Their standing feud has been described in John Mather's book *The Very First Light*. So I will not fault Smoot for not looking critically over Mather's shoulder to determine for himself if the discovery of the FIRAS blackbody was shipshape. But Smoot did have an obligation (and the expertise) to do so before proclaiming the success of his own experiment.

When Smoot began to look at his own data from the DMR experiment, he saw two clear problems: From his sum channels he saw that there was not enough power in the sky corresponding to a 2.7 K blackbody. And his orbital determination of antenna patterns showed in all cases that something was clearly amiss. This was an additional proof that there was no blackbody – and Smoot would surely have recognized this.

But to come out and say so would mean giving up on discovering the Big Bang Baby Universe.

So Smoot seems to have done two things: He produced his own (false) corroboration of the Mather blackbody. And he put the problem with his antenna patterns in the backburner, promising to come back to them. He never did. He received his Nobel Prize while this Sword of Damocles was hanging over him.

If we accept that George Smoot was supremely competent at what he did, there is no escaping the conclusion that he knew in the 1990-1992 timeframe that there was no blackbody in the sky, and so his skymap had no connection to any Baby Universe.

And to think that Smoot was nearly going to get away with it. The Sword of Damocles was going to be suspended eternally over him, secured so it would never fall.

CHAPTER V-4
Herbert Gush

Herbert Gush is at first a tragic hero. He was on track to be the first to make the discovery that John Mather made, and so perhaps to win the Nobel Prize. Gush seems to have been a passionate and quiet researcher who worked with his students and colleagues like a close-knit family. He worked under tight funding conditions. It seemed fitting that the heavens would anoint such a man. But that was not to be.

Some mishap with his experiment beyond his control set him back by a few months, and he made his discovery within a couple of months of John Mather. His results were just as impressive as those of Mather. Certainly, Mather's Nobel Prize candidacy was strengthened by the Gush result because of the phenomenal corroboration of a fantastically accurate discovery.

The Nobel Foundation, however, did not include him in the 2006 award even though there was room for a third Laureate. It may be that Gush's candidacy was never put forward.

Herbert Gush's discovery of the blackbody was erroneous. But this can be classified among errors many discoverers make in the course of a normal scientific process. It may be that the Mather discovery somehow made Gush feel overconfident. I have not seen anything to suggest that there was foul play in Gush's case.

A retired Herbert Gush – who today reportedly lives in Italy and has the vocation or the avocation of a farmer – has reasons to be thankful. Better to be a retired and contented and respected scientist than a Nobel Laureate whose discovery turns out to be bogus or scammed. As he enjoys his life in Italy, Gush should be grateful that providence guided him out of harm's way. He remains the rare honorable man of Big Bang cosmology.

CHAPTER V-5
Saul Perlmutter

I have described the Saul Perlmutter discovery in Chapter II-5, and in Chapter IV-4 I have presented the scientific fallacy of his discovery. As to his overall scam, it is a very intricate one and is not easy to describe to a general audience. So I have devised a way to explain it in terms of what I will call the Big Bang Firewall.

Figure V-1 (*facing page*):
Cartoon illustrating the Perlmutter Discovery Template
(a) Saul Perlmutter plotted astronomical observations of intensity *vs* redshift on a piece of paper. These included the older data of 'nearby' objects (solid circles) and his and his colleagues' new data on 'distant' objects (open circles).
(b) He then constructed a template of distance *vs* velocity purely from Big Bang theory.
(c) Then he made a whole host of assumptions – each questionable – to convert this template to an intensity *vs* redshift template that can be compared with the measurement data.
In this template, the central line represents an expanding Universe which is largely empty. The upper line (broken line) represents a Universe with conjectural Dark Energy. The lower line (dotted line) represents a Universe with conjectural Dark Matter. Dark Energy would push out and accelerate the expansion, and Dark Matter would gravitationally pull in and decelerate the expansion.
(d) By overlaying the template of (c) on the observational data of (a), Perlmutter found that his distant cluster fell on the upper line. And so he reported discovering *both* the Accelerating Universe and Dark Energy.

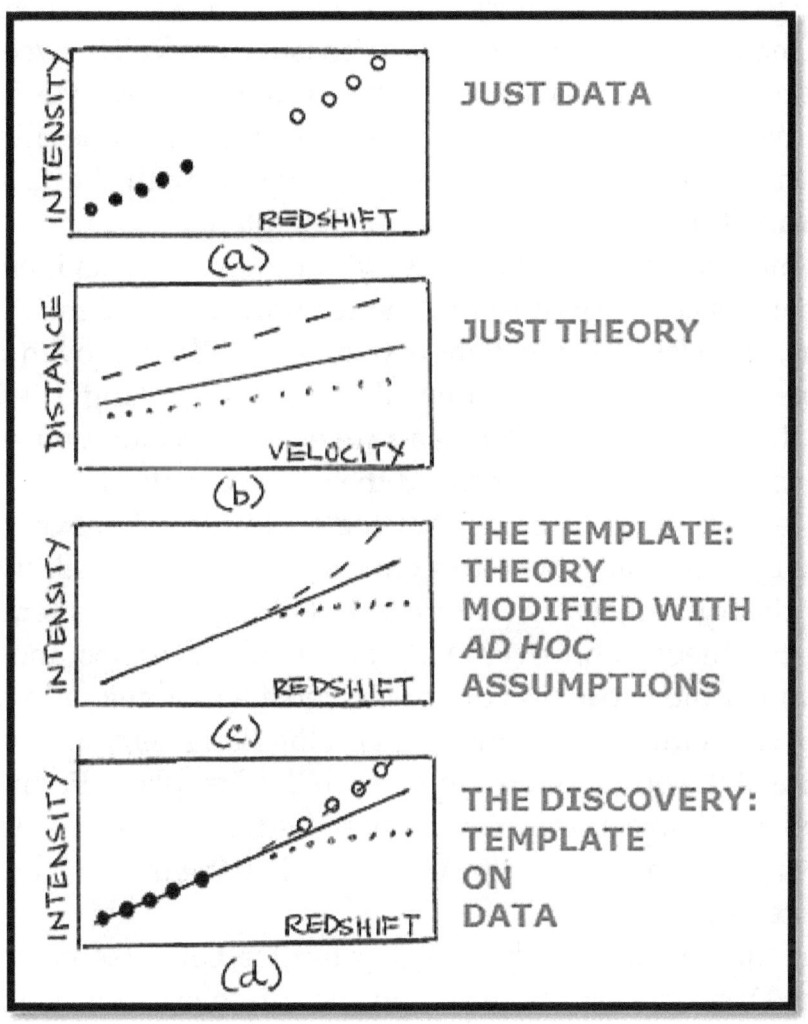

Figure V-1

Any astute scientific expert who has watched closely the simultaneous, seemingly synergistic, development of Big Bang cosmology on the ground-based astronomy front (iconized as Saul Perlmutter) and the satellite-based observation front (iconized as John Mather) would have noticed that there has been maintained most carefully an invisible firewall between the two arenas. No side acknowledges that their discovery is in any way *contingent on* the other side's discovery. If any one side falls flat on its face, the other side will carry on with the mission, unharmed. That seems to have been the plan.

Thus, for instance, *when* (not *if*) all the satellite evidence of Big Bang is accepted at last by the scientific establishment to be bogus, the classical Big Bang expansion will be proven to be bogus. Here is where the firewall will save Big Bang. It will be said that since Perlmutter has proved the acceleration of the expansion, he has tacitly proved the expansion itself.

Next we need to recall that the Perlmutter discovery was made by combining his astronomical observations with the Big Bang theory framework (Figure V-1). He constructed a template from the theory and overlaid it on his observations and thus concluded the expansion of the Universe was accelerating. The discovery is thus 100% contingent on Big Bang being the correct science of the origin of the Universe in the first place. Perlmutter added the conjectural Dark Matter and Dark Energy to the Universe over and above the Classical Big Bang theory, and thus executed his discovery.

In the end it was that the Perlmutter Universe contained some Dark Matter as well. So he discovered a Universe containing a mix of both, their proportions having been pinned down also.

Now, what is the support for this theory framework that makes the template a <u>scientific fiduciary</u> with which discoveries can be made? Satellite observations. In particular, the 2.7 K Big Bang blackbody and the Baby Universe skymaps. But the skymaps are contingent on the blackbody. So the entire template is contingent on the blackbody.

Note that the 'nearby' cluster of data in Figure V-1 cannot by itself support the fanned-out portion of the Template unless the presence of Dark Matter and Dark Energy were independently proven by observation.

Presumably, Perlmutter's adding the conjectural Dark Energy and Dark Matter to the Universe and accelerating it thus does not retroactively change the 2.7 K Big Bang relic blackbody prediction, which has been clinched to phenomenal accuracy. There is no going back and revising *that* number to suit the Perlmutter expansion, should it predict a different relic radiation temperature because of the existence Dark Matter and Dark Energy that was not figured in the original framework.

So either Perlmutter has to accept the 2.7 K and fall with it when it falls; or he has to reject the 2.7 K and lose his Discovery Template. Either way, he has no discovery of any kind. Just some fine astronomical observations.

Thus it cannot be said that there is a Big Bang Firewall between the Perlmutter and the Mather discoveries or between their scams. As goes Mather so goes Perlmutter.

> You can't dance at two weddings with one behind.
>
> Yiddish proverb

CHAPTER V-6
National Aeronautics and Space Administration

V-6.1 Speculative scenarios

I will describe below what happened within NASA following my exposing the COBE satellite fraud in April 2007. The cause-and-effect connection between my exposé and what happened subsequently in NASA is entirely speculative. I have no direct knowledge that anything happened within NASA *as a result of* the exposé. But I will try to be reasonable in my speculation.

V-6.2 The capping of John Mather's NASA rank

In Appendix A, I have described in detail how John Mather, who had been elevated in April 2007 to the high position of NASA Headquarters Chief Scientist in recognition of his Nobel Prize, was sent back by then NASA Administrator Michael Griffin to Mather's pre-Nobel Prize post in March 2008. This post is that of a Senior Astrophysicist in the Observational Cosmology Laboratory of the Goddard Space Flight Center, with the additional task of the Senior Project Scientist for the James Webb Space Telescope project.

In the nearly seven years since then, a lot has happened with Mather in terms of awards and honors and coming his way. These have been touted in NASA official announcements. Photos of Mather with powerful members of the United States Congress have also been bandied about. Mather has stood on the same stage with the United States Secretary of Defense Robert Gates. Mather was declared by TIME Magazine as one of the world's twenty-five most influential people in space, and NASA gloried over it. He was showcased by National Geographic and Smithsonian Institution. He gave two

Commencement Speeches. And it has been amply clear that high NASA officials such as Christopher Scolese have Mather's back.

Michael Griffin, Charles Bolden, and Christopher Scolese

But in spite of this monumental prominence in the world stage, John Mather's rank remains capped at his exact pre-Nobel Prize level. It is almost as though someone much higher up has drawn a line that is not to be crossed, no matter what. Even when Christopher Scolese succeeded Griffin as Acting NASA Administrator and even when the new Administrator Charles Bolden took over, no changes in Mather's aforementioned job status occurred. Nor was Christopher Scolese able to change this situation when he became the Goddard Center Director.

In my thinking, when Griffin took the drastic step of sending Mather the American superhero back to his pre-Nobel Prize job rank, he must have felt the need to consult the White House (which supervised Griffin.) So this matter was agreed upon there. It was probably felt that this was all the action the Government could take. The rest was up to the scientific establishment. But the scientific establishment never did its part. At any rate, because the decision was made in the White House, no one could undo this.

V-6.3 Cronies keep Mather in constant limelight

Within the above constraints, there then appeared to be a much heightened sense of urgency in Goddard to keep Mather in constant limelight. It was as though there was a concerted plan to celebrate Mather out of the allegation of fraud.

John Mather seems to have built a huge base of loyal friendships throughout his career. He would advise young people to "talk to people", that everything good he has done came from talking to people. These friendships were now paying off. A great many people who surround Mather at Goddard are also his Facebook Friends.

The James Webb Space Telescope is a high visibility platform, and NASA has a high visibility website. On the right hand side of that website there used to be then a news column that was constantly updated. For a time, that column was kept abuzz with John Mather news. It seemed as that if he sneezed, it would be news. If he coughed, it would be news. Such was the extent of the grooming there. In time someone in NASA took care to dampen this activity.

A colleague of Mather at Goddard Space Flight Center has a daughter who, at the time I speak of, was studying at the University of Maryland. She was the president of the student body there and as such, had a strong say on who was to become the commencement speaker that year. Following her father's suggestion, she contacted Mather who immediately accepted. So Mather came to be the commencement speaker. Outwardly, it appeared that the university leadership had chosen him after full-court deliberation, ignoring the public allegation of fraud.

There were various video and audio clips made of Mather by NASA and released. One bizarre plan went like this: The (audio) interview had the make-believe situation that this was the Oscar Award Day. Celebrities were arriving, and reporters were corralling them in the lobby to do a quick interview. John Mather was as such a celebrity arriving.

V-6.4 The name-branding programs

But what takes the cake in the bizarre category is the setting up of the John Mather Nobel Scholar Program at Goddard in the summer of 2008 by a private organization called the Henry Foundation. These internships were open only to Goddard summer interns. The chosen interns would be granted the right to be called John Mather Nobel Scholar, and awarded a travel grant of $3000. The Henry Foundation was later replaced by the National Space Grant Association as the host of this program.

This was followed in its wake in 2009 by the setting up of the John Mather Congressional Policy Internship program by the American Institute of Physics. It seems that gifts from Mather come with strings attached. The candidates must agree to be branded forever with his name.

Technically, none of these is an official NASA Goddard activity. But in spirit they were very much Goddard dos. Goddard after all was Mather's identity, and that is the home he was operating from. For a long time, the line between these branding activities and NASA was kept blurred. The John Mather Nobel Scholars and the John Mather Congressional Policy Interns were sometimes brought together for purposes of tours etc.

But in the summer of 2013 when Goddard was under the directorship of Christopher Scolese, a big push was made to make the John Mather Nobel Scholar Program seem like a Goddard in-house program. Shortly before the application deadline for the program, John Mather gave a grand lecture to the all the summer interns in a large Goddard hall, making expansive hand gestures that were expertly photographed and published. There it was said or hinted that the John Mather Nobel Scholarship is a pathway to a regular civil service job at Goddard. Two NASA announcements were put out about this, one titled "Once a Mather Scholar, always a Mather Scholar." This was followed by the announcement that a large crop of eleven interns had received the scholarship.

V-6.5 Albert Einstein World Award of Science

When Charles Bolden took over from Michael Griffin as NASA Administrator in the summer of 2009, he inherited the John Mather problem. From the events that unfolded since, it appeared that he was content to maintain status quo in this matter. He did not change anything about Mather's job situation. There seem to be no public photographs of him with the most famous scientist of the organization he was to lead. Nor are any public accounts of the two meeting each other.

But in September of 2013 I came upon a letter by Bolden – dated November 30, 2012 – recommending Mather for a major international award: The Albert Einstein World Award of Science. The content of this letter was most astounding. There was no sign that NASA was aware of any problems with Mather. Instead, he was presented as one of the greatest figures in science for all time. This shattered all conjectures I had that Mather had secretly faced some accountability within NASA. It became unclear to me now if Michael Griffin had indeed taken any punitive action at all on Mather.

Normally, nomination letters such as this (Figure V-2) are kept confidential. People hear about the awards when they are announced. In this instance the letter was placed on an Internet site – probably for private access by reviewers of the nominations. But the site was not a private access site. This is how I came upon it during a keyword search.

So is it possible that Bolden felt that he was writing a confidential nomination? What I noted about the letter is its unpolished language and grammar. Clearly, it did not go through many desks as such organizational letters do before they are finalized for signature. It seems to have been drafted by one person and sent straight to Bolden. That person logically might be Mather's backer Christopher Scolese, Goddard Center Director.

The letter thus raises to my mind many disturbing questions about NASA. It seems now that the NASA cover up of the fraud is far more aggressive than I thought.

In the end the nomination did not succeed. The award was given very aptly to the renowned medical scientist Paul Nurse, also a Nobel Laureate.

V-6.6 Summary

The decision to cap Mather's job – if that is what it was – was a right thing to do. It showed leadership responsibility. However, this was only a first step of more steps that needed to follow. That did not happen.

Goddard scientists have behaved badly, and seemed to have lost sight of the fact that they were on public payroll. They were occupying positions of public trust. It was their place to get to bottom of the allegation. For they, more than anyone else, knew what went down at Goddard with the COBE satellite. Instead, they engaged in what amounts to a collective cover up with purposeful deliberation.

Figure V-2 (*following two pages*): Charles Bolden nominates John Mather for the Albert Einstein World Award of Science.

November 30, 2012

Lillyan Hernandez
Secretary General
Consejo Cultural Mundial
Apartado Postal 10.1083
Col. Lomas de Chapultepec,
C.P. 11002 Mexico, D.F.
MEXICO

Dear Ms. Hernandez:

Thank you for the invitation to nominate candidates for the 2013 World Awards in Science and Arts. On behalf of the National Aeronautics and Space Administration (NASA), it is my pleasure to submit the nomination of John C. Mather, NASA Goddard Space Flight Center, for the 2013 Albert Einstein World Award of Science.

Dr. Mather is an eminent astrophysicist and cosmologist whose work and leadership has led to major advances in the understanding of our origins and place in the universe. His signature work, for which he was awarded the Nobel Prize in Physics in 2006, was in pioneering observations of the relic background radiation left over from the Big Bang, which led to the resolution of fundamental questions on the nature of the universe. Currently, Dr. Mather is serving as the scientific lead for the James Webb Space Telescope (JWST), which when launched in 2018 will be the largest, most complex telescope ever launched into space. This technological marvel is designed to observe the formation of the first stars and galaxies a few hundred million years after the Big Bang and the atmospheres of Earth-like planets orbiting around other stars. It will provide a true leap forward in our ability to understand our origins and search for life beyond our solar system. The concept for the JWST mission was developed under Dr. Mather's leadership beginning in 1995, and he continues to provide top-level scientific leadership on every aspect of the project both nationally and internationally.

In addition to the Nobel Prize, Dr. Mather's achievements have been recognized with membership in the National Academy of Sciences and with numerous and wide-ranging awards. He is the author of many publications, including his book entitled "Very First Light," which was written along with John Boslough. He has given many public lectures to help the public understand his work and to inspire young people to be as excited about science as he has been. In 2007, Mather was listed among Time magazine's 100 Most Influential People in The World and most recently in 2011 listed among Time magazine's 25 most influential people in space.

Figure V-2 – Part 1 of 2

286

Dr. Mather's outstanding work and leadership has provided a unique service by opening a window into space and avenues of discovery for both scientists and the public to access to a deeper understanding of space and astronomy and a higher appreciation for our place in the universe. Please accept this nomination of John Mather for the Albert Einstein World Award.

Sincerely,

Charles F. Bolden, Jr.
Administrator

Enclosures

Figure V-2 – Part 2 of 2

CHAPTER V-7
The physics establishment

V-7.1 The Doyens: Steven Weinberg and Stephen Hawking

After I exposed the COBE satellite fraud in the spring of 2007, after NASA capped John Mather's rank at the pre-Nobel Prize level in the spring of 2008, and after the airing in the summer of 2010 the PBS-NOVA program *Hunting the Edge of Space* that surgically expunged John Mather and George Smoot from the history of cosmological observations, one might have expected some honest voices from within the physics established to arise – whether in assent or dissent or concern. That has not happened to this day in the winter of 2015.

wikipedia.org

Steven Weinberg, Stephen Hawking, P. J. E. "Jim" Peebles, and Michael Turner

Among the living Big Bang cosmologists, Steven Weinberg and Stephen W. Hawking are certainly the foremost in terms of their actual scientific involvement and leadership, and also in terms of their public visibility and stature. Not only have they assumed stewardship of Big Bang cosmology but also their popularizing of this subject has garnered them personal wealth. In this pursuit they continue to be active today. These are two exemplary establishment figures to look at, first and foremost.

Steven Weinberg of the University of Texas at Austin, a Nobel Laureate in particle physics, has written the iconic book *The First Three Minutes* which has become a symbolic tome of Big Bang cosmology to the science-loving world citizenry. And it is clear that until recently he has remained active in research in Big Bang cosmology. In the face of stark developments in evidence in public, he has remained entirely undeterred. There is no diminishing in his zeal.

To some, Weinberg's support of Big Bang cosmology may have a particular ring of scientific authenticity. There exists today a view that Big Bang cosmology has religious overtones or undertones, that there is a religious agenda behind it. That may have had the potential of diminishing the subject as a pure science. But Weinberg is a self-proclaimed atheist. So he strengthens Big Bang in a way no one else does.

In December 2014 Weinberg gave a lecture at Harvard University that had the flavor of the distilled wisdom of a long and distinguished career in physics. He said that the quest in particle physics should be directed a very high energies (~ 10 trillion times the highest energy we can harness now) and very small size scales (~10^{-17} to 10^{-19} nuclear radii.) Where would one find such large energies? According to a report on his lecture:

But the experiment may have already been done, by nature, and there may be a way to look back at it, Weinberg said. During the inflationary period immediately after the Big Bang there was that kind of energy, he said, and it would be evident as gravitation waves in the cosmic microwave background, an echo of the Big Bang that astronomers study for hints of the early universe.

Thus Steven Weinberg's scientific worldview is informed by imagined events of fourteen billion years ago. In his mindscape the Big Bang inflationary space has gradually morphed to as hard a metal-and-mortar surrounded reality as the core space of a particle accelerator. This should give us deep pause.

Stephen Hawking of Cambridge University is in constant evidence as the most populist promoter of Big Bang cosmology. He certainly holds the loudest loudhailer: When Stephen Hawking speaks, the world hears. His description of the COBE satellite discovery of the picture of the Baby Universe as "discovery of the century, if not all time" not only reverberates around the world but had also been used by the Nobel Prize-giving body to justify the prize. Less known is his applauding of the WMAP satellite's "evidence for inflation" as the most exciting development in physics during his career. And in 1999, by changing his mind on the acceleration of the expansion of the Universe (from being against it to being in favor) and declaring that it is "very reasonable" that there should be Dark Energy, he emphatically supported that discovery. To this day he has not taken the tiniest step back from all these positions.

Certainly, no one expects Stephen Hawking to be up on all that is going on in his field. It may be that he gets selected scientific briefings from his support group. But when Hawking issues sweeping opinions on current developments in Big Bang cosmology and commands instant worldwide attention, one expects that he would be kept briefed also on the problems thereto pertaining. His pronouncements are not just something to fascinate the world but they actually have tangible effects that are beneficial to some and not so to others. I have already explained how his opinion carries weight with the Nobel-givers. This also means Hawking helps drive out other cosmologies.

Stephen Hawking's commitment to Big Bang was reaffirmed in an interview he gave to the WIRED Magazine, reported on 1 January 2015. He was asked what people need to understand to keep up with the current state of cosmology. His reply:

They need to   understand that the Universe began with a period of inflation in which it expanded at an ever-increasing rate. Quantum fluctuations would have caused some regions to have expanded slower than the rest of the Universe. These regions would eventually stop expanding and collapse to form galaxies, stars and all the structure in the Universe. Quantum fluctuations during inflation would also generate primordial gravitational waves.

This vivid description of the events of the inflation era reminded me of a little boy who, just after his first visit to Krispy Kreme, most animatedly described how donuts are made on the conveyor belt – from bulk dough to rows upon rows of the glazed product – in endless variety.

No other person living today has used his position of public trust to push a totally bogus field of science as has Stephen W. Hawking. No other scientist living today has so much authoritative opinion to offer concerning a field (space technology) he knows absolutely nothing about. He has no business endorsing before the world complex experiments after such experiments have been called into question by an expert.

As a famous quotation wrongly ascribed to him has it: "The greatest enemy of knowledge is not ignorance, it is the illusion of knowledge."

To be deeply ensconced in the imagery of altruistic nobleness of the academia and to make so much hard cash off such bogus science – this must be a technique unique to Big Bang cosmology and its sister field String Theory.

Those who seek to influence the course of our civilization (in this case, science) using the enabling platform they have been given by the society must also assume corresponding accountability. This accountability may not be suspended for anyone. No one is exempt from this. Not even a poignant genius.

V-7.2 The Flamekeepers: P. J. E. Peebles and Michael S. Turner

P. J. E. "Jim" Peebles of Princeton University is a major figure in Big Bang cosmology and indeed, in its history. A worthy disciple of the late great Robert H. Dicke, he was the theoretician to complement Dicke the engineer. Here is Peebles in his own words in an interview conducted by Institute of International Studies, University of California, Berkeley. He is speaking of the Big Bang relic radiation:

After the war, my teacher, Bob Dickie, was the one who said, "What is this radiation? We ought to look for it." ... we should just remind ourselves that it is radiation that fills space, that pretty unambiguously was produced when the universe was young, has all the properties of thermal radiation, and that it could only have acquired those properties when the universe was young and dense and hot. He said to me, "Go think about the theoretical considerations." That set my career.

And that career was to take him to the acme of the Big Bang cosmology establishment. It was thus most fitting that he would write an endorsement for the John Mather epic *The Very First Light*, the greatest adventure of all scientific adventures:

The very best and most edifying description I have seen of how to do big science.

But if Peebles had unstinting praise for those who supported Big Bang, he also had scathing criticism for those who opposed it. Here is what he said in an *American Institute of Physics* interview about Hannes Alfvén (a leading Big Bang dissident):

I had some pretty harsh words to say about it (Alfvén's theory) because I thought there was no way you could understand the very close to perfect isotropy of the thermal radiation in such a universe.

And of course, because of Big Bang's phenomenal publicity machine, the world believed him and others like him in this regard, and came to the see the Big Bang critics as some kind of nuisance for the progress of science. It certainly did not help that Alfvén was a publicity-averse introvert.

Within the academia today, Michael S. Turner of the University of Chicago might be identified as Big Bang's most indefatigable popularizer, a worthy successor of George Gamow in that capacity. From his academic platforms to such high worldwide media platforms as *Scientific American*, he is very much in evidence with his constantly updated recipe for the Universe (x% regular matter, y% Dark Matter, ... etc.) With his resounding assurances to the world that he and his colleagues are closing in on the Universe, he has been a soothing voice of hope and comfort to a Big Bang-wowed world.

So far-reaching is Turner's outreach that I once saw the spectacularly colorful Big Bang cover of *Scientific American* in a busy grocery store chain, right on the same rack as the tabloid magazines near the check stands. The line in front of me was long enough that I could thumb through the pages in a leisurely manner and be wowed by the scientific authority and authenticity the article exuded.

Michael S. Turner has also been a father figure to the BICEP2 team, and has helped calm the troubled waters when the team faced criticism as to whether their discovery concerned Big Bang gravitational waves or merely interstellar dust. As the head of the Kavli Institute of Cosmological Physics at the University of Chicago, he is well-connected to the Kavli Foundation and has almost certainly helped BICEP2 get funding from that source.

V-7.3 The great endorsers: Edward Witten and David Gross

And then there is Edward Witten of Princeton University, promoted by many as the foremost theoretical physicist today, and also the smartest man on the planet. Normally he would not be a subject of discussion here because this String Theorist's involvement in Big Bang cosmology has been peripheral at most. But when the news of the Nobel Prize for the discovery of the acceleration of the Big Bang expansion of the Universe broke in early October of 2011, Witten came out with a sterling endorsement. He said something to the effect that it took (even) him a considerable amount of time to cogitate before he saw the beauty of the idea of Dark Energy (something the 2011 Nobel Laureates purportedly discovered.) So here was Edward Witten putting his considerable scientific clout behind Big Bang, in the fall of 2011.

wikipedia.org

Edward Witten and David Gross

Personally, my criticism of Witten is tinged with a bit of respect: He has not sought to make money off science the way his Big Bang counterparts have, although copious amounts of prize money have spontaneously come his way. This point is important because, in the last analysis, Big Bang is about money.

Significantly, when the BICEP2 matter created tremendous stir in the scientific world, Witten remained silent. Wisely so.

David Gross received the Nobel Prize for Physics in 2004 for his work in particle physics. He then expanded his interest to String Theory and Big Bang cosmology. He in fact became a very vocal patron of both fields, and in that capacity, gave much positive media exposure to these fields. This is what he told Television's NOVA program about Big Bang:

And that explosion or hot state left remnants that we can observe today in the microwave background. So we know that that aspect of the theory is true.

And since the Big Bang explosion did take place, he concluded, we need a new theory to understand the earliest times of that explosion. That is why, he said, we need String Theory.

So Gross is in fact pushing not just Big Bang theory or String Theory, but he is pushing them together as interrelated sciences. With the power of his Nobel Prize, he is leveraging one bad science on another bad science.

David Gross has continued zealously in his campaign to promote the fraudulent 2.7 K Big Bang blackbody. In the Lindau Nobel Laureates Meetings in July 2012 there was a discussion session involving Gross, Robert Laughlin (a physics Nobel Laureate) and some others. Gross forcefully averred:

We have to recognize that microwave background is the most perfect ever measured blackbody radiation – ever – any laboratory on Earth …

When Laughlin expressed doubt, Gross said:

You wanna make a little bet …?

The wager was made but Gross was never heard from again.

V-7.4 The Fundraiser: Roger Blandford

On 6 December 2011, Stanford University physics professor Roger Blandford testified before the US House of Representatives, *Committee on Science, Space and Technology*. He was a witness in the hearings on the James Webb Space Telescope (JWST), very much a Big Bang cosmology program (although, with Big Bang in trouble, attempts are currently underway to spin it as an exoplanet-hunting telescope.) He was appearing in his capacity of Chairman of the Astronomy and Astrophysics Decadal Survey (Astro2010; Figure V-3), a committee that would recommend what research funds should spent on for the next decade. This committee is centrally responsible for the billions of dollars wasted on the bogus science of Big Bang. This committee had all the information available on the COBE satellite fraud, and as such it was incumbent on them to put screeching brakes on Big Bang. Instead they floored the gas pedal.

wikipedia.org

Roger Blandford

Roger Blandford testified under oath:

JWST is specialized to observe in the infrared region of the spectrum. This is relevant because, although much light emitted by the most distant galaxies is in the optical and ultraviolet spectral bands, the wavelengths of this light are stretched roughly tenfold through the expansion of the Universe into the infrared band, as we push out to greater distance and earlier times.

With the COBE satellite evidence gone, there was no evidence whatsoever of stretching of space. Nor was there any support for the following statement Blandford placed on sworn record in order to garner large amounts of public funds for Big Bang cosmology:

We now have a fairly precise *standard model* of cosmology, which allows us to predict the approximate date when the first stars and galaxies formed. This lies well within JWST's reach and it will be able to observe the resulting *redshifted* optical and ultraviolet light. It will help explain just how the gas in the Universe was converted from atomic to ionized form during the so-called *Epoch of reionization* which marked the end of our cosmic *dark age*.

Figure V-3 (*overleaf*): *Astro2010*
The Astronomy and Astrophysics Decadal Survey:
This is a partial list of names of people responsible for wasting billions upon billions of American tax dollars on the bogus science of Big Bang cosmology.

Survey Committee Membership

Roger Blandford, Chair, Stanford University
Lynne Hillenbrand, Executive Officer, California Institute of Technology
Martha P. Haynes, Vice Chair – Science Frontiers, Cornell University
John P. Huchra, Vice Chair – State of the Profession, Harvard-University
Marcia J. Rieke, Vice Chair – Program Prioritization, University of Arizona
Steven J. Battel, Battel Engineering
Lars Bildsten, University of California, Santa Barbara
John E. Carlstrom, The University of Chicago
Debra M. Elmegreen, Vassar College
Joshua Frieman, Fermi National Accelerator Laboratory
Fiona A. Harrison, California Institute of Technology
Timothy M. Heckman, Johns Hopkins University
Robert C. Kennicutt, Jr., University of Cambridge
Jonathan I. Lunine, University of Arizona
Claire E. Max, University of California, Santa Cruz
Dan McCammon, University of Wisconsin-Madison
Steven M. Ritz, SCIPP, University of California, Santa Cruz
Juri Toomre, University of Colorado at Boulder
Scott D. Tremaine, Institute for Advanced Study
Michael S. Turner, The University of Chicago
Neil de Grasse Tyson, American Museum of Natural History
Paul A. Vanden Bout, National Radio Astronomy Observatory
A. Thomas Young, Lockheed Martin Corporation [Retired]

Panel on Cosmology and Fundamental Physics (CFP) Membership

David Spergel, Chair, Princeton University
David Weinberg, Vice Chair, Ohio State University
Rachel Bean, Cornell University
Neil Cornish, Montana State University
Jonathan Feng, University of California at Irvine
Alex Filippenko, University of California at Berkeley
Wick Haxton, NAS, University of Washington
Marc Kamionkowski, Caltech
Lisa Randall, Harvard University
Eun-Suk Seo, University of Maryland
David Tytler, University of California at San Diego
Clifford Will, Washington University

Figure V-3

V-7.5 The anointers: C. Megan Urry and Andrew Baden

C. Megan Urry is a well-known astronomer and a professor at Yale University. She is a stalwart figure in her field. In 2009, well into the period John Mather's discovery had been placed in question, she invited Mather to deliver a distinguished lecture at Yale. There, she praised him for his "selfless" service. In 2012, as an elected official of the American Association for the Advancement of Science (AAAS), she seems to have seen to it that Mather became a Fellow of that high intelligentsia.

wikipedia.org

C. Megan Urry, Andrew Baden, and Yervant Terzian

But of course Megan Urry is by no means alone in having John Mather's back – as the colloquialism goes. Many academics all over the nation were instrumental in inviting Mather to their institutions to be feted in a high profile manner, often covered in the media. Many more were the moving force behind his getting awards and accolades. Here is one more academic who was in evidence in this regard.

Andrew Baden is the Head of the University of Maryland's Department of Physics where John Mather has long been an Adjunct Professor. On Baden's watch, good things happened there to John Mather. He was appointed to the prestigious position of College Park Professor – considered a full faculty slot. There was developed an outreach program to connect to high schools and Mather was chosen to be an inspirer to schoolchildren. This fact was then bandied around. In the spring of 2013, Mather gave the high profile commencement speech to the University's College of Computer, Mathematical, and Natural Sciences. Earlier in 2008, Mather had also given the Maryland commencement speech.

And then there is Yervant Terzian, a distinguished professor of astronomy at Cornell University. When a Lecture Series named after him was endowed, Mather was the inaugural speaker. Later, as a leader of the National Space Grant Association, Terzian would agree for that respectable organization to become the host the John Mather Nobel Scholar program.

V-7.6 The Berkeley incubator

Big Bang cosmology has been nurtured in no place as it has been nurtured in the University of California at Berkeley (UCB) and the Lawrence Berkeley National Laboratory (LBL). For our purpose these two can be considered as a single organization energetically seeding, growing and then proliferating Big Bang cosmology. Of the seven Big Bang Nobel Laureates, three are intimately connected to this organization.

John Mather was a doctoral student at UCB. He was nurtured by his teacher Paul Richards, a big Berkeley Big Bang cosmologist. What Mather learned from Richards would shape the fateful NASA COBE satellite project. But later the United States Department of Energy also would lay their claim to fame for supporting graduate student Mather through a grant at LBL.

George Smoot was a long-standing postdoctoral researcher at LBL, not only in connection with the COBE satellite but also before that in connection with doing Big Bang cosmology from balloon-borne and aircraft-borne instruments. After his Nobel

Prize in 2006, he would establish his own premier center for research in Big Bang cosmology at UCB.

Saul Perlmutter did his doctoral research at UCB under the guidance of Richard Muller, a UCB doctorate himself. After receiving his Ph. D., Perlmutter was employed by LBL.

After their charges got the Nobel Prizes and after the allegation against them was in profuse evidence, UCB and LBL continued to fete them proudly and loudly. The now famous Berkeley Nobel Laureate Parking Permits were handed out and streets of the LBL campus were named after Smoot and Perlmutter (see chapter II-7.)

In 2014 there emerged on the Big Bang discovery scene yet another Berkeley cosmology Ph. D., Chao-Lin Kuo, working at Stanford University. He was the designer of the novel electromagnetic instrumentation for the BICEP2 telescope that discovered the B-mode polarization swirls in the sky, indicative of inflationary gravitational waves. That discovery was disputed by members of the scientific establishment. It was established that while the instrument did discover B-mode polarization in the sky, this feature was due to interstellar dust and not gravitational waves.

I have maintained however, that there were no B-mode polarization swirls in the sky and that what the BICEP2 telescope saw was an instrumental artifact.

If an institution is credited for nurturing a body of students that go forth in the world and do good things, the institution should also be discredited when the trainees do bad things. For Berkeley, the latter circumstance applies.

Other organizations deeply involved in promoting experimental or observational Big Bang cosmology include the Johns Hopkins University, Princeton University and Stanford University.

V-7.7 Professional Societies: *American Institute of Physics*

The American Institute of Physics (AIP) is the most central and the most comprehensive professional society in physics, not just in the United States but in the whole world. It is for the most part a bureaucratic umbrella for physics and physics-related professional societies. Its social arm – one that holds meetings and gives out awards and accolades etc. – would be the American Physical Society (APS). The AIP has never been known to promote any single branch of physics especially or to promote any single individual especially.

wikipedia.org

Frederick Dylla and Louis J. Lanzerotti

In 2006, H. Frederick Dylla became the Executive Director of AIP. Somewhere along the line he and John Mather latched on to each other. In 2009 the AIP established the Mather Congressional Policy Intern program, to be hosted jointly by the AIP and the United States Congress. The program was funded in part from Mather's Nobel Prize money and in part from the proceeds from selling a house Mather had donated to AIP. Each summer this internship program would place two young students deep in the recesses of the United States Congress. This was the beginning of a powerful relationship between Dylla and Mather.

From here on, John Mather was always visible in the AIP

setting, whether it was to hobnob with his own summer interns or to inspire all the AIP interns as a group. He was then inducted to a committee to oversee donations to AIP. He was in evidence in little parties given by or connected to AIP. He was present in photo-ops with the AIP interns in the Congress. Mather's involvement with the interns then extended even to inviting the interns' parents to lunch at AIP at the "Mather Table". Dylla would also use the AIP subsidiary Student Physics Society (SPS) to give Mather further exposure in meetings and publications. Suffice it to say, in the face of allegations of science fraud, Dylla went out of the traditional AIP way to single out one physicist to keep him in such limelight as would cast aspersions on the physicist's scientific critics. It was then said that AIP was hobnobbing so much with Mather especially because he was an accessible Nobel Laureate.

All of these may seem quite innocuous and even positive. After all, Dylla is using his personal friendship with Mather to benefit the AIP. Who in his right mind can criticize this noble act? But here is the real issue: Dylla and those who supervise the position he holds (represented by Louis J. Lanzerotti) were aware of the two high profile public issues about Mather, raised in 2008 (confinement of Mather by NASA to his pre-Nobel job rank) and 2010 (expunging of Smoot and Mather from the history of cosmology in the PBS-NOVA program *Hunting the Edge of Space*). It was incumbent upon them to look into this before instituting/continuing a program that would in effect brand young people with the Mather name. And here is the most important point: These AIP leaders were the most conveniently situated people to get the best possible advice on the allegations against Mather. For some of these people, such advice was available just down the hallway or reached by a short walk through the campus. For others it was just a phone call away.

But they did not seek out such advice, or if they did, they got bad advice. They proceeded most resolutely to institute Mather as an AIP icon, and thus as an American physics icon.

The AIP Mather Congressional Policy Interns and the John Mather Nobel Scholars would naturally sing the paeans of John Mather in the social media. Sometimes their home institutions would put out laudatory news items about the internships. This way Mather would be kept in limelight as a shining beacon of inspiration to the young of America. The world at large – which gets its information from the mass media – would not have a clue that there was anything wrong with this picture.

V-7.8 The blackbody die-hards

By 2014 much of the Big Bang establishment had gone silent on the COBE 2.7 K blackbody – but not the die-hards. They continued resolutely to teach and preach this subject to the public. They either did not want to know about, or did not understand the John Mather fraud.

columbia.edu   stsci.edu

Amber Miller and Mario Livio

Of a number of such scientists, two latest examples of such assertions are from Amber Miller of Columbia University and Mario Livio of the Space Telescope Science Institute. They are practicing scientists whose government-funded research rides on the false assumption that the blackbody is there in the sky. They are also posited as pillars of the Big Bang community. In addition, Miller is an academic leader and Livio is a science popularizer.

In the summer of 2014 Miller appeared in a panel discussion at the World Science Festival. There she seemed to make it a point to mention the phenomenally accurate blackbody discovered by John Mather. No other panelists took any issue with that.

In December 2014 Livio – writing a column in *Physics Today* with Marc Kamionkowski – averred that CMB is a blackbody.

These and other scientists like them seem outwardly as the odd people who never got the word – like the two World War II Japanese soldiers hiding from the Allied Forces in jungles of Mindanao until 2005. In truth, what the scientists are doing may be quite deliberate.

V-7.9 Awards and accolades

The scientific field of Big Bang cosmology is especially rich in having a large supply of awards with which to anoint its members. These include within-establishment and outside-establishment awards. The former are the standard accolades of the physics establishment such as the numerous professional society prizes. The latter include the Gruber Prize (endowed by Peter Gruber), the Kavli Prize (Fred Kavli), the Milner Prize (Uri Milner), the Shaw Prize (Run Run Shaw), and others.

During the period that the field of Big Bang satellite discovery as a whole was placed in question, the above awards and accolades kept being showered on the experimenters and the theorists alike. The Gruber Prize, for example, was awarded to Charles Bennett, John Mather, Saul Perlmutter and Brian Schmidt. The Shaw Prize was awarded to Charles Bennett. The

Milner Prize was awarded to Big Bang theorists Alan Guth, Stephen Hawking, and Andrei Linde. The Kavli Prize was awarded to Alan Guth and Andrei Linde. These prizes in effect repeatedly injected steroids into an enterprise that should have been allowed to die a natural death.

The illustrious Paris Observatory, a most respectable and long-traditioned institution, has been particularly active in feting Nobel Laureates John Mather and George Smoot. Both have been given the Observatory's prestigious Chalonge Medal, and invited to give speeches.

In the fall of 2013, Charles Bennett was awarded the Jansky Prize of the National Science Foundation, given by the National Radio Astronomy Observatory.

V-7.10 The lecture circuit, the festival circuit

The Big Bang Nobel Laureates have been ubiquitous on the lecture circuit – both academic and general. The allegation against Big Bang had no effect on this activity. The hosts included many universities – large, medium and small – and many other organizations. Thus the rosters of many distinguished and long-standing lectureships are today adorned with the names of the Big Bang Nobel Laureates.

There stands out one particular lecture by John Mather. In 2009 he was invited to present a talk at the meeting of the Mars Society. It is not known if the organizers had asked the invited speakers to bring their own introducers (BYOI?). But Mather did. What stands out is that Mather brought an attorney. The introducer introduced himself as a NASA Legal Advisor, and then proceeded to introduce Mather. What stands out even more is that he said that Mather's accepting the Nobel Prize while being a government employee has been certified to be appropriate. One wondered what that was all about. Was

Mather lawyering up? Was he trying to convey to someone a veiled cease-and-desist notice – at government's expense?

The festival circuit refers to the burgeoning phenomenon where somebody sets up a science or technology "festival", ostensibly to inspire the young. Two such forums in the United States are the World Science Festival and the USA Science and Engineering Festival. Both have embraced Mather in a big way, and do not miss a chance to showcase him.

wikipedia.org

Brian Greene and Lawrence Krauss

The World Science Festival was created by, and is now run by Columbia University physics professor Brian Greene. It is a platform that combines science and performing arts and cooking and some other things. It attracts great many people and receives coverage in the media. So for John Mather to be adopted for exposure by this organization is quite a powerful thing.

Arizona State University professor Lawrence Krauss, a flamboyant science popularizer, is very much in evidence at the World Science Festival. In the summer of 2012 he staged a big show starring a number of big name Big Bang cosmologists, including John Mather. They were all seated on stage in a neat crescent. When Mather's turn came to speak, he said that while a great many papers had been written on the COBE satellite discovery of the Baby Universe map, not a single paper was written on his discovery of the blackbody spectrum. Presumably he meant critical papers. Upon this, Krauss nodded his approval. The other Big Bang cosmologists, by their presence, indicated their acquiescence.

Then in 2013, in an article he wrote in the *New York Times* on the findings of the Planck satellite, Krauss once again emphatically mentioned the blackbody as clinched science.

In the same timeframe as the above feting, there started burgeoning in India 'techfest' conferences that went by such names as Techniche and Cognizance. The organizers of these found a good thing in John Mather. They would invite him and he would accept to give a videoconference talk from the US. This way, the organizers got a Nobel Laureate speaker (and a sahib to boot) without paying the prohibitive international travel expenses – quite a feather in their cap. This activity just went on and on. For a while there John Mather became a prominent fixture in the Indian techfest circuit.

If the Indian festival organizers were making a beeline for John Mather, the Government of India was no slouch. In January 2010, the Government of India, on the recommendation of the Indian Space Research Organization, conferred on John Mather the President General Gold Medal, and hosted him at the Indian Science Congress. Mather was then feted in various forums and featured prominently in the Indian media. He gave assurances to the Indian masses that the world would not end in 2012 – as the Mayan calendar held. In December 2013 again, he was invited to a "Conclave" by the Indian Institute of Information Technology, Allahabad, supported by the Government of India.

Next to the United States, India is on record as the nation most active in feting John Mather as an inspiration to her young.

V-7.11 The commencement circuit: *University of Maryland* and *Notre Dame University*

The University of Maryland, it seems, just cannot get enough of John Mather. Not only did they make him the Commencement Speaker in 2008, but they invited him back to deliver the spring 2013 commencement address for graduates of the College of Computer, Mathematical, and Natural Sciences at the University.

In spring of 2011, Notre Dame University conferred on Mather the honorary doctorate degree. He was honored in this way alongside Robert M. Gates, then US Secretary of Defense and Shirin Ebadi, the Iranian Peace Nobel Laureate in 2003, and other luminaries. Here is what the citation said, in part:

He and George Smoot received the Nobel Prize for their development of the Cosmic Background Explorer satellite project that transformed the study of the early Universe from a largely theoretical pursuit into direct observation and measurement.

Here it is important to keep in mind that Mather's selection had to begin in the Notre Dame's Department of Physics where people had the full capacity to understand the allegation of fraud. They advanced Mather's candidacy in the face of this existing situation.

V-7.12 Big Bang Public Outreach Program

Big Bang cosmologist George Gamow, who was also a pioneering science popularizer, may have been the first visionary in seeing the crucial need of, and in endowing his field with, a robust Public Relations program – the kind that is today known as Public Outreach. In plain language, Gamow saw the media as an integral component in developing the science of Big

Bang. Today, no single field of physics gets as much media exposure as Big Bang. The copious supply of fine art-like images of Big Bang Universes are an enormous asset in this context.

The continuous stream of Big Bang bloviation emanating from the so-called science reporters inundates the world today. The platforms are as high as TIME Magazine, National Geographic, Smithsonian Institution, Scientific American, Public Broadcasting Service and so on. No sooner did John Kovac spin his BICEP2 B-mode swirls than he became one of TIME's Most Influential Person in the World. Not to be outdone, the journal *Nature* anointed BICEP2-buster David Spergel as one of "10 people who mattered this year."

And do not forget the big publishing houses either, putting out one bestselling Big Bang tome after another. Many such books adorn the airport bookstores. A popular television series also helps. What other field of science has such global Technicolor presence? Why is it that the most unscientific of all scientific ventures is also the most loudly touted?

But the most effective public outreach today may be Big Bang's Wikipedia sites. They reach out to the most relevant segment of the public: the science-savvy citizens. They in turn spread the good word to the citizenry at large.

If one studies these Wikipedia sites which cover every single sub-topic of Big Bang cosmology, one quickly realizes that they exude quality and excellence. They have been written in depth and in breadth. They have been written meticulously and professionally. In short, they have been written anonymously by Big Bang insiders, or by people in contact with Big Bang insiders. This is a well thought out, concerted Big Bang stratagem without appearing to be so.

In 2014 Deepak Chopra, sporting a new look and new couture by some fashion designer no doubt, came out swinging with a new television gig called *The Future of God*. This was replete with references to Big Bang, connecting it to such things as mind, consciousness, the source and some other such sublime things. It seemed that Chopra was trying to put his platform on scientific pillars, and Big Bang was one of his major pillars. It seemed that he had boned up on the subject well. And it seemed that in the end, everything is inside each human being. Big Bang in the end is within each of us.

What better platform than a Deepak Chopra gig to spread the word from? Who has a greater reach among the masses? Not even Stephen Hawking. Big Bang may finally have found its true resting home. It certainly needs one.

V-7.13 The Big Bang cosmology establishment

This entire book is about the Big Bang cosmologists and so it may seem strange to have a subsection on this subject. And yet, there is one point that needs to be brought to laser focus because it is the key to understanding what this community truly is.

For this purpose take me out of the equation so that we have an unbiased field. Assume that none of the issues I have raised exists. Then posit two self-evident questions:

1. When the WMAP satellite released its results in 2009, why did not this establishment ask the first, the foremost, the most logical and the most crucial question: Where is the five-frequency WMAP CMB spectrum?

2. The same question about the nine-frequency Planck satellite CMB spectrum (2013).

Why have they never asked these questions to this day?

There is only one explanation: They all privately know that the above two satellites have found that their blackbody is not there in the sky. The only way to continue their enterprise was to maintain collective silence on this.

So dishonesty comes very easily to the Big Bang establishment. It has become their second nature.

And of course we must now ask a third question: Why did the physics establishment at large say or do anything about this?

Had the obvious question been asked in 2009, the Big Bang framework would have unraveled and the Perlmutter-Riess-Schmidt Nobel Prize would not have gone through. The American physics establishment did a number on the Nobel-givers.

A larger, historical perspective on the Big Bang establishment is out of the scope of this book. But I will discuss one aspect by way of giving the reader a taste of what that perspective consists of. This one concerns the clever and repeated employment of mass psychology techniques by the establishment over a period of nearly a century.

What is *mass psychology*? It is what the New York advertising agencies employ to help corporations sell their products. It is what Chairman Mao used to transform China. But in this art the Big Bang cosmologists are the granddaddy of them all.

There have been four major applications of special mass psychology techniques by Big Bang cosmologists to date:

1920s: The Hubble expanding Universe diagram – *derived* from some underlying observations under various assumptions and hypotheses – was installed in the public mind as itself an *observational* diagram.

1960s: The observed Microwave Background Radiation in the sky was quickly dubbed *Cosmic* Microwave Background Radiation. Then the distinction between it and a bogus *relic radiation* predicted from Big Bang cosmology was gradually erased from the public consciousness. The two became one and the same.

1990s: The *acceleration* of the expansion of the Universe was touted to high heaven, creating the subliminal sense in the public mind that the expansion itself was a foregone conclusion.

2000s: They introduced the nomenclature CMB Angular Power Spectrum to describe the distribution of the intensity fluctuations in the skymap. This was a strange nomenclature to use. But its purpose became clear when the expression was gradually shortened to CMB Power Spectrum, and thence to just CMB Spectrum. The stratagem here is to convey to the unsuspecting public that the CMB Spectrum is a fully clinched thing, where the public is expected to think that the CMB Intensity Spectrum is a clinched thing. That is, they are deep-sixing the fact that the blackbody is not there in the sky and then letting people think that it is there – through this subterfuge.

In modern times, hairy mathematics (e.g., inflation theory) has been used as the organ grinder's monkey. Expensive space-borne technology has been used ineptly just to say to the public: "We're hip! We got moves!"

Of course I am not suggesting that the Big Bang cosmologists hold well-organized planning meetings in some kind of a Star Chamber and develop and implement such strategies. These things happen naturally because of an ingrained culture that has developed over time. They think alike and act alike and this is how we get the impression of methodical planning. Deception comes naturally to them.

313

CHAPTER V-8
The peer review system

V-8.1 The peer-review issue

Peer-reviewed publication of a discovery is the first and the foremost requirement for a Nobel Prize in physics.

A discussion of the merits and the defects of the present scientific peer review system – even if confined to physics – is a large one that can fill an entire tome. For our purposes, only a very specific topic needs to be addressed: How was it that the bogus Big Bang discoveries got published in prominent peer-reviewed journals over a long period of time, and subsequently stood unassailed, leading up to the Nobel Prize and beyond?

V-8.2 Nature of astrophysical publications

There are two special aspects of astrophysical publications that we need to be aware of.

First, research papers in astronomy, astrophysics and cosmology – such as published in *Astrophysical Journal* for example – are necessary allowed a great degree of laxity as compared to physics papers. In the former, order-of-magnitude estimates and simplified modeling assumptions are the norm. But when these are invoked, the final conclusions of the papers are also understood to be correspondingly lax.

Big Bang cosmologists have exploited this beneficial laxity. Assumptions and estimates are made to arrive at a result. In a later paper this result is invoked with a citation to the earlier paper, but the result is now treated as clinched science. This overt scam works because the peer-review system has allowed it to work. Both John Mather and Saul Perlmutter have used this subterfuge to arrive at their discoveries.

Second, when astrophysical papers involve techniques or

technologies outside the normal range of referee expertise of the journals, evidence shows that such issues are not properly evaluated by seeking out proper expert referees. Basically, the authors are trusted to have done things right. And yet, by publishing these papers, the journals provide the body of evidence necessary to clinch a discovery. This has happened with the Penzias-Wilson discovery as well as the Smoot-Mather discovery. As to the Perlmutter discovery papers, the special reviewer needed would have to have been a trained logician, beyond being an expert scientist in the field. Such was the depth of the intricate scam.

V-8.3 How the peer-reviewed publications came about

Let us take the three Nobel Prize-winning Big Bang discoveries one by one.

The Penzias and Wilson discovery

The Penzias-Wilson discovery was made in 1964 and the papers were published in 1965. Robert Dicke of Princeton University convinced Penzias and Wilson that the single frequency observation they had made was one point on the Big Bang relic blackbody spectrum; that Penzias and Wilson had actually discovered the blackbody spectrum.

This would at first sound quite unlikely. But that aspect was cleverly brushed under the rug when two back-to-back papers appeared in *Astrophysical Journal*. Dicke and some Princeton colleagues wrote a paper setting the stage for a ~ 3 K blackbody radiation spectrum to be found in the sky. Penzias and Wilson reported in a paper immediately following in the same issue of the journal that they found precisely that blackbody. Nobody saw the oddity in going from one observation point to an entire spectrum, when in fact that one observation had not even been shown to fall on that spectrum.

So this bogus discovery (see Figure V-4) came about as a result of:

(a) Dicke's manipulation of Penzias and Wilson;
(b) Referee error or referee collusion;
(c) Editor error or editor collusion; and
(d) Scientific establishment's acquiescence.

As to the last point, it remains an abiding mystery why the radio astronomers did not spot the gaping fallacy. It was entirely within their expertise to do so. It was in fact bread-and-butter science for them.

Another point to keep in mind is that Dicke was himself an engineering expert on what Penzias and Wilson did. He had to know what he was doing.

The Smoot and Mather discoveries

The main discovery papers of George Smoot and John Mather were published in *Astrophysical Journal*. Figure V-5 shows some of these papers that were cited by the Nobel Committee in making the award.

These authors did publish some of their papers in engineering journals, thus exposing their claims to true experts. Most of these publications were on small pieces of their project and did not contain the issue of scientific defect in the overall discovery. So this seeming transparency was in fact irrelevant.

Measurement of Cosmic Microwave Background Radiation
Penzias, A.A.
IEEE Transactions on Microwave Theory and Techniques
vol.mtt-16, no.9 p.606-9
Publication Date: Sept. 1968 Country of Publication: USA

Intergalactic H I Absorption at 21 Centimeters.
Penzias, A.A. ; Scott III, E.H.
Bell Telephone Labs., Holmdel, N. J.
Astrophysical Journal, 153: L7-9(July 1968).

Isotropy of Cosmic Backround Radiation at 4080 Megahertz.
Wilson, R.W.; Penzias, A.A.
Bell Telephone Labs., Holmdel, N. J.
Science, 156: 1100-1(May 26, 1967).

Determination of the Microwave Spectrum of Galactic Radiation.
Penzias, A.A.; Wilson, R.W.
Bell Telephone Labs., Inc., Holmdel, N. J.
Astrophysical Journal 146: 666-9(Dec. 1966).

Measurement of the Flux Density of CAS A at 4080 Mc/s. (Flux Density of
Cas A Measured, Using Horn- Reflector Antenna, Response Compared with
Output of Reference-Noise Source)
Penzias, A. A.; Wilson, R. W. /Bell Telephone Labs., Inc., Crawford Hill
Lab., Holmdel, N.J./.
Astrophysical Journal, Vol 142, Oct. 1, 1965, P. 1149-1155.

A Measurement of Excess Antenna Temperature at 4080 mc/s. (Effective
Zenith Noise Temperature of Horn- Reflector Antenna at 4080 mc Due to
Cosmic Black Body Radiation, Atmospheric Aborption, etc)
Penzias, A. A.; Wilson, R. W. /Bell Telephone Labs., Inc., Holmdel, N.J./.
Astrophysical Journal, Vol 142, Jul. 1, 1965, P. 419-421.

Figure V-4: Some of the initial publications of Penzias and Wilson
(Source: Bell Laboratories). Notice that all these engineering papers
were published in *Astrophysical Journal*. When Penzias finally
published a paper in an engineering journal, it was in the nature of a
review paper.
The Penzias-Wilson discovery never received a timely engineering
review which would have revealed the fallacy of making an entire
spectrum out of a single-frequency measurement.

There was, however, one key paper by J. Mather, M. Toral and H. Hemmati titled "Heat trap with flare as a multimode antenna" that was published in *Applied Optics* in 1986. This paper contained the proof-of-concept study for FIRAS optics. So there the crux of the matter was indeed presented to the right evaluators.

A reading of this paper suggests that objections were raised by referees and the authors were asked to respond. Mather's response was simply to agree that there were aspects to his experiment that remained questionable. With these *mea culpas* duly recorded, the paper got published. So the journal and the referees at least rose to the occasion.

But now came one of Mather's clever twists that he is so good at. Like an eel he wiggled out of the situation and proceeded to his discovery. He declared (just *declared*) that his instrument had been tested to be in perfect order and ready to launch. So if appropriate referees provided some safeguards, with Mather no safeguard was of any avail.

Smoot, G. et al. 1990 Astrophys. J 380, 685
Smoot, G. et al. 1992 Astrophys. J (Letters) 396, 1

Mather, J.C. 1982, Opt. Eng. 21, 769
Mather, J.C. et al. 1990, Astrophys. J. (Letter) 354, 37
Mather, J.C. 1994, Astrophys. J. 420, 440
Mather, J.C. et al. 1999, Astrophys. J 512, 511

Figure V-5: Some key Smoot-Mather papers cited by the 2006 Nobel Committee in their award description.

Astrophysical Journal now provided the clinching publications, its referees never bothering to check the all-important Mather-Toral-Hemmati paper.

With the Saul Perlmutter discovery, *Astrophysical Journal* was certainly the right place to publish the results (Figure V-6). All subjects covered in the paper were within the purview of such a journal. There is no problem in this area.

The problem is sweeping things under the rug: Passing off approximate ideas and speculative results of yesteryear as clinched science, and then building clinched discoveries upon them. This is basically like a building a monstrous house of cards by simply scotch-taping each joint. This is what the discoverers did and the journal blessed.

---

[27] A.G. Riess et al., "Observational evidence from supernovae for an accelerating universe and a cosmological constant", Astron. J., **116**, 1009-1038, (1998),

[28] S. Perlmutter et al., "Measurement of Ω and Λ from 42 high-redshift supernovae", Astrophys. J., **517**, 565-586, (1999),

[29] A.G. Riess et al., "Type Ia supernova discoveries at $z > 1$ from the Hubble Space Telecope: Evidence for past deceleration and constraints on dark energy evolution", Astrophys. J., **607**, 665-687, (2004),

---

Figure V-6: Key papers cited by the Nobel Committee in awarding the 2011 Nobel Prize for physics.

V-8.4 *Physical Review Letters*

The publication of a paper in *Physical Review Letters* is undoubtedly the ultimate imprimatur of quality in the physicsdom. This status has come about from a long tradition of publishing seminal papers. But tradition is one thing. What the editors and referees do at a given time is another.

The journal published Herbert Gush's discovery of the blackbody spectrum in 1990. Clearly, there existed around the submission great atmospherics. John Mather had just reported his discovery, beating out Herbert Gush by mere months. Gush's discovery results were said to be even more precision than Mather's. All these created a compulsion for the journal to publish the paper, once it was submitted. The editors and referees probably would not have taken it upon themselves to raise any serious objections at that juncture.

And yet, the paper should have been stopped, for reasons I have explained in Section IV-2. But to invoke those reasons, the referees needed to have the right expertise. It is not clear that appropriate expertise was brought to bear on the paper.

Likewise, there also existed great atmospherics around the BICEP2 discovery when that paper was being refereed at this journal. A discovery of great import had been reported with great fanfare. The discovery was then called into question with great fanfare. A compromise was then reached, and it was reflected in the manuscript. The referees may have imposed some revisions. The paper was then published.

If the manuscript had been disassociated from the atmospherics and referees with appropriate expertise were consulted, the paper would have been stopped. The editors would have to have displayed the courage to reject the paper, causing perhaps much consternation in the physics establishment.

But none of that happened. The paper sailed through fine. And this publication enshrined the first discovery of B-mode polarization swirls in the sky.

These are two stark examples of a journal that passes as the toughest place to publish, publishing totally bogus discoveries. These examples prove that the journal is as good as its editors and referees. A highly respectable journal can become a disreputable journal based merely on how these people conduct the process.

V-8.5 Peer review and the funding mechanism

Publishing is not the only arena in which the peer-review system has been corrupted. The funding process in Big Bang cosmology – as in other fields of science – also revolves around peer review and peer patronage. These processes go by names such as proposal review panels, decadal surveys and so on. Conducting these processes legitimizes in the public's eyes the transfer of moneys from the treasury to the Big Bang cosmologists.

And yet, these processes in Big Bang today are as corrupt as the corresponding process in publishing. When the scientists deep-six inconvenient results, the Government funding agency grant monitors go along. So who is watching? Who is challenging? Absolutely no one. When a group of nobles stand together and lie in unison, what recourse does anyone have?

V-8.6 Chapter conclusion

Astrophysical Journal, more than any other journals, systematically helped create a false body of evidence that would result in the award of a series of Nobel Prizes. This cannot be happenstance. This was a clear act of commission.

The decadal surveys have acted as enablers of the Big Bang cosmologists. By corrupting a good custom, these peer groups have endorsed one fanciful notion after another as genius science the nation must urgently support at any cost. The scientific establishment at large has aided and abetted by their silence.

CHAPTER V-9
The United States Congress

During the period of my observation – the time following my exposing of the John Mather NASA fraud in the spring of 2007 – there have been numerous publicized contacts between Mather and members of the United States Congress. These fall in two general categories: contacts through Mather's home institution, NASA Goddard Space Flight Center; and contacts through Mather AIP Congressional Policy Intern program.

wikipedia.org

Representative Bill Foster, Representative Rush Holt, and Senator Barbara Mikulski

The first member of the Congress that comes to mind is Representative Bill Foster, a Democrat from Illinois – a suave gentleman with an impressive presence. He used to be a physicist before he turned to politics, and thus may have been known to Mather and Frederick Dylla, leader of AIP. By today it has become almost a routine that Foster would appear each summer for a photo op in the Capitol Hill with Mather and the two interns for that year. Sometimes there would be a lunch for the four in the Congressional Dining Hall. These photographs and accounts of the lunch by the interns (invariably singing paeans of Mather and Foster) would be bandied around in AIP websites. These are powerful images to tell the world that Mather moves in the hallowed corridors of American power.

And a minor footnote that may have nothing to do with anything: Over a period of years, Mather has contributed thousands of dollars of his own money to the Foster political campaign fund.

Representative Rush Holt, a Democrat from New Jersey and a former physicist, also hosted a Mather intern. Frederick Dylla is on record as a subsequent contributor to his campaign.

The next person that comes to mind is Senator Barbara Mikulski, a Democrat from Maryland. Of the multiple contacts between her and Mather, one that stands out for me is where the two are sitting on a sofa facing each other, in a living room type setting. Another has Goddard Center Director Christopher Scolese, Sen. Mikulski, two astronauts and John Mather. The Senator is looking at Mather, smiling. This photo was published in a Congressional website. In another all-hands meeting in Goddard, the Senator – reading from a cue sheet – makes certain to mention " … our very own in-house Nobel Prize winner…". And so it goes.

There are photos of Mather with a number of other very powerful members of the Congress. There are enough lawmakers involved for a generalization to be made that Mather is strongly connected to the United States Congress – at least the Democratic side of it.

Whether intentionally or unintentionally, these reports create the image of a scientist whose back the United States Congress has, even as he stands accused of fraud and even as this is known to the leaders of NASA and AIP. The Congress itself also stands informed of this allegation through its official channels. There is a message being broadcast by the United

States Congress to anyone seeking to expose John Mather's NASA COBE satellite fraud: He is under our protection.

In Appendix C, I describe how the Mather matter was referred to two Congressional bodies having direct and immediate oversight authority over NASA. These committees were then headed by Senator Bill Nelson, a Florida Democrat, and Representative Bart Gordon, a Tennessee Democrat. They had websites inviting public comments. So my representation was submitted following their prescribed rules.

wikipedia.org

Senator Bill Nelson and Representative Bart Gordon

Now, what could rise to a higher level of oversight concern than the allegation that the national space agency was committing the grandest possible science fraud and deceiving the entire world? Almost anything, it seems. There is no indication that any attention was paid to this matter. Is it any wonder John Mather felt as safe in his endeavor as he did?

Corruptio optimi pessima
(Corruption of the best is the worst of all.)

Latin phrase

CHAPTER V-10
The White House

V-10.1 Why the White House?

John Mather is a Civil Service scientist working for an organization that reports directly to the President of the United States. For a time, he was a NASA Chief Scientist who reported to a NASA Associated Administrator who reported to the NASA Administrator who in turn reported directly to the President. So Mather was in fact only a couple of steps below the highest office of the land.

In routine matters, it seems that the NASA Administrator interacts with the Office of Science and Technology Policy (OSTP) in the White House, rather than directly seeing the President. My earliest referral of the science fraud was to the OSTP, in 2007. Around this time I had also referred the matter to NASA HQ and NASA-OIG (Appendix C).

I have no knowledge of what became of these referrals or which, if any one, was acted on. But in the spring of 2008, the then NASA Administrator Michael Griffin oversaw the abrupt and not-too-amicable departure of John Mather from the Office of Chief Scientist back to his old job at Goddard (Appendix A).

My thought at the time was that Griffin would have consulted the OSTP before taking such a drastic step about an iconic and high level American Government scientist. The fact that no subsequent NASA leaders could reverse that decision to keep Mather confined to his pre-Nobel job status also suggests that the decision was supported at a higher level than the Administrator. Even the very strong support Mather enjoyed from powerful members of the United States Congress did not change this situation.

V-10.2 The referral to President Barack Obama

After a multi-year struggle to bring the NASA COBE satellite fraud to public account, I had finally referred the matter to President Obama (see Appendix C.) I never received a response. This matter concerns a public referral to a public office. Therefore, as a citizen, I have every right to engage in my own speculation – based on visible facts – as to what transpired.

wikipedia.org

President Barack Obama, John Holdren, and Carl Wieman

My letter concerned a matter of science. It is thus most likely to have gone to the White House science section (OSTP). My initial thinking was that it would have gone to the White House Science Advisor John Holdren (Director of OSTP) for possible forwarding to the Office of the President. I did not know that the White House has a Chief of Staff for Science. It now seems to me that that is where my letter most logically went.

The Chief of Staff for science then was James Kohlenberger. This is what I have learned about him since: The man has little qualification in science. But he has the backing of former US Vice President Al Gore. That may be how he came to power in the Obama White House. Kohlenberger was deeply involved with NASA in his official capacity, and as such had dealings with NASA Deputy Administrator Lori Garver.

My letter not only criticized NASA, but also specifically the Deputy Administrator.

Following my letter of 24 August 2010, on 18 October 2010 the principal in the COBE satellite fraud – Nobel Laureate John Mather of NASA Goddard Space Flight Center – was feted in a White House science event. I saw a connection between the two, and took it to be a pointed public snub. And also a resounding affirmation of John Mather by the highest office in the land – *after* an allegation was brought to its attention as a last resort.

For a moment I thought I had come to the end of the line. But only for a moment. This situation did not make any sense to me, given what President Obama stood for.

If good sense and good judgment were the rule in President Obama's office, there was no way this would have happened with the knowledge of that office. The only other alternative was that the letter was intercepted and deep-sixed.

I made a big stink in my various Internet sites. What is this?! I bring an allegation to the President about a cover-up and feting of science fraud by some of the most powerful people and groups of people in the nation, and the President takes right over and continues with the very same cover-up and feting in the very same manner! What is this?!

It seems to me that this stink somehow reached an appropriate person in the White House who was unaware that this was going on in the name of the President. In February of 2011, a very high level change took place in the White House, very quietly. The media seems to have remained surprisingly silent about this – as if they did not want to touch this one. Only a few blog sites carried it.

In February of 2011 the high-flying James Kohlenberger, a person of little science qualification cozily installed in the dizzying height of power as the White House Chief of Staff for

Science – without having lined up other gainful employment and without his White House successor named – abruptly left.

Unrelated to this matter, Lori Garver left her coveted NASA position in September 2013.

V-10.3 John Mather in the White House

While this appeared to be the extent of what transpired in the White House, I was later to find out that there had been a number of formal visits by John Mather to the OSTP following my referral (Figure V-7). His host at the White House was Carl Wieman, Associate Director of Science at the OSTP (and a Nobel Laureate in physics.)

There was a meeting between Carl Wieman and John Mather on 7 October 2010. It was followed by a meeting on 7 January 2011 that was attended by a number of officials. After that there was a lunch hour meeting between Wieman and Mather on 20 July 2011.

Nothing more is known. John Mather continued to be feted unabated in public as a celebrity NASA Nobel Laureate. The Universe-sized scientific construct based on his discovery has garnered more Nobel Prizes – just because no one made a determination on the allegation of fraud and thus raised issues with the overall science.

Carl Wieman left his position in the White House in June 2012 for health reasons.

If the above meetings in the White House were the result of my referral, it remains unclear what came of them. It is therefore difficult to know if to praise or to fault the White House on its role, if any, in this matter.

However, high level White House officials seem to have spent much time on my letter. That would deserve resounding praise. That would be activity in accordance with excellence.

DETAILS

Visitor	John C Mather
Host	Carl Wieman
Total People	1
Access Type	VA

APPOINTMENT INFO

Appointment Number	U47254
Appointment Start Date	10/7/2010 14:30
Appointment End Date	10/7/2010 23:59
Meeting Location	NEOB
Meeting Room	5224

SCHEDULING INFO

Appointment Made Date	10/4/2010 15:46
Caller	Donna Coleman
Officer that Entered Appointment	YT
Computer Used to Enter Appointment	WIN
Officer who Last Updated Record	YT
Most Recent Update to Appointment	10/4/2010 15:46

DETAILS

Visitor	John C Mather
Host	Carl Wieman
Time of Arrival	1/7/2011 13:52
Time of Departure	1/7/2011 16:07
Total People	8
Access Type	VA

APPOINTMENT INFO

Appointment Number	U73017
Appointment Start Date	1/7/2011 14:00
Appointment End Date	1/7/2011 23:59
Meeting Location	NEOB
Meeting Room	5235A
Post of Arrival	K101
Post of Departure	K1

SCHEDULING INFO

Appointment Made Date	1/6/2011 13:01
Caller	Donna Coleman
Badge Number	79147
Officer that Entered Appointment	A5
Computer Used to Enter Appointment	WIN
Officer who Last Updated Record	A5
Most Recent Update to Appointment	1/6/2011 13:01

DETAILS

Visitor	John C Mather
Host	Carl Wieman
Time of Arrival	7/20/2011 11:32
Time of Departure	7/20/2011 13:24
Total People	2
Access Type	VA

APPOINTMENT INFO

Appointment Number	U27759
Appointment Start Date	7/20/2011 12:00
Appointment End Date	7/20/2011 23:59
Meeting Location	NEOB
Meeting Room	5235
Post of Arrival	K101
Post of Departure	K1

SCHEDULING INFO

Appointment Made Date	7/18/2011 0:00
Caller	GREGORY GERSHUNY
Badge Number	87026
Officer that Entered Appointment	GG
Computer Used to Enter Appointment	WIN
Officer who Last Updated Record	GG
Most Recent Update to Appointment	7/18/2011 17:14

Figure V-7: White House visitor logs of John C. Mather visiting with members of the Office of Science and Technology Policy.

CHAPTER V-11
The Nobel-givers

V-11.1 The Seven Pillars of Big Bang

The seven Big Bang cosmology Nobel Laureates had of course received many other high profile prizes and accolades by the time they stood at the threshold of the House of Nobel. Of all these prizes and accolades I will discuss the Nobel Prize especially here not because it is the biggest prize of them all; but because – in the present context – the Nobel Prizes have been an enabler while the other prizes have been the anointers.

For more than three decades, the relationship between Big Bang cosmology and the institution of the Nobel Prize for Physics has been symbiotic. They have nourished each other.

Every time a Big Bang Nobel Laureate lecture is staged, the hosts do not miss the opportunity to make the most of the Nobel name. All of the copious and relentless Big Bang media exposures – in print or broadcast – make the most of the Nobel name. To many honest experts in the scientific field, Big Bang today has surely emerged as bad science. But it is thriving in the public arena because of the Nobel fortifications. Big Bang cosmology today stands tall on seven Nobel pillars.

The Nobel name in physics has been bandied around in the modern world by Big Bang cosmologists like it has not been bandied around by any other groups in physics. The glitter on television, the color extravaganza on Google image pages, the popular science magazines and the books and the DVDs galore – all these constantly keep the Nobel-Big Bang nexus in evidence before the world citizenry lest they forget for a moment. Anyone, especially anyone outside the establishment, who critiques Nobel-winning science is surely seen as a pesky nuisance, and perhaps worse.

V-11.2 The Nobel-givers

As it is well-known, the Nobel Prize for Physics is decided by the membership of the Royal Swedish Academy of Sciences. The actual work of receiving the nominations and evaluating them is done by the six-member Nobel Prize Committee for Physics. In the end, however, the entire Academy votes on the recommendation put to them by the Committee. I do not know if this is just a rubber stamp, or if the Academy members exercise individual judgment.

However, the Academy should not be seen in isolation from the scientific establishment at large in an us-*vs*-them sense. The Academy is a part of this establishment and depends on the latter to provide the Nobel candidates pool, to nominate them and to help evaluate them.

So the truth about the dynamics of this relationship lies somewhere between two extremes. The first one is where the Nobel-givers hold all the power, and the second one is where the establishment dictates to the prize-givers. A science historian can probably use the Nobel deliberations archives released to date to make some conclusions in this regard. But this is beyond me. I will only address the matter of Big Bang cosmology in an effort to determine where the truth lies in that context.

The first point I will make is based on personal experience. I have known some Swedish scientists, to varying degrees. I have also seen some at work in their own institutional setting in Sweden. It is very hard for me to accept that the outside scientific establishment can manipulate or dupe them – in the conventional sense. The Swedish physicists are nobody's fools. So the second extreme above was unlikely to have prevailed in the case of the Big Bang Nobel Prizes.

The second point is that the three sets Big Bang Nobel Prizes thus far span more than three decades. There has been a change of generation in the Royal Swedish Academy. In the case of each set of Big Bang Nobel Prizes, some fifteen years passed after the discovery and before the prizes were given. There was no cutting corners, no rush to judgment. So these awards are not in any way aberrations or deviations from the norm. They are the result of the full-court evaluative procedure. They were done by the book.

The third point is that I do not know (and cannot find out) if the Nobel evaluation process includes a fresh evaluation of the science *ab initio*. Since the subject is taken up for consideration by the Nobel-givers only after a long period of being vetted and debated within the scientific establishment, it is possible that such a fresh evaluation – if it does take place – is only cursory and not adversarial.

All these points give rise to a most puzzling issue: How could patently bogus discoveries have been anointed again and yet once again? The fault with these discoveries was as evident when the awards were made as it is today.

V-11.3 It is not the science, it is the atmospherics

So my theory – and a theory is all it is – is that there existed or was created around each discovery great atmospherics and great theatrics that worked on the subconscious minds of the Nobel evaluators (along with everyone else.) They were unable to separate what the great Magus on the stage wanted them to see and what he did not want them to see. The Magus can keep every single pair of eyes of the thousands of pairs – even very sharp eyes – focused on the waving wand in his right hand while his left hand is doing something else.

Let me take the three sets of Big Bang Nobel Prizes one by one. By the very nature of the subject (theatrics and atmospherics), I must of necessity be a little verbose in discussing it.

Penzias and Wilson 1978

There had existed for a long time the most portentous observational discovery of Edwin Hubble, and it cried out for some type of spectacular acknowledgement on the world stage (even though Hubble was long gone.) The Big Bang cosmologists paved the way for this by showing that the Hubble discovery was the key to the origin of the Universe.

So it was nothing less than the birth of the Universe that was being addressed. It was like no other matter in science. The Nobel Prize needed to break out of narrow boundaries and prove worthy of the great deeds of man that were unfolding. The Big Bang cosmologists were setting the course of the civilization in the high frontier.

They also predicted – quite non-intuitively, very self-assuredly and at the great risk of being proved wrong – that the Universe today would be filled entirely with a continuum of an ethereal placenta from its birth (actually, the relic radiation; but the subliminal cuts are very important part of the created atmospherics.) It was also a rigorous scientific prediction because it was experimentally falsifiable.

George Gamow and Fred Hoyle – both pioneers in science popularizing – were fighting it out in the open over Big Bang, whipping up great public interest. Stephen Hawking later kept up this interest quite admirably. The world was ready for this crescendo to reach some type of splendiferous climax.

In the middle of all this came two unassuming engineers blundering across the sky with their antenna. They were baffled at what they saw. To make a long story short, unbeknownst to themselves Arno Penzias and Robert Wilson had fulfilled the highly improbable prediction the Big Bang theorists had made. A truth that is the strangest of all fictions! No less a refereed platform than the storied *Astrophysical Journal* published the back-to-back papers, one setting the stage for the discovery and the other springing the discovery. The correctness of the discovery was thus placed beyond question, and the legitimacy of the discovery was placed beyond reproach. Cute stories were now hung on the discoverers for public consumption, involving bird guano and double barrel shotgun and such.

Is it any wonder that in the fullness of time the Nobel Prize would crown this indomitable advancement of our knowledge of the home we inhabit? What kind of a scientist would stand in the way?

But if some hard-nosed Nobel evaluators had any lingering doubts, these were allayed from the most unexpected of quarters. The two stalwarts among the Big Bang critics, Hannes Alfvén and Fred Hoyle, came forward with theories that explained the Penzias-Wilson blackbody radiation in terms other than Big Bang. In so doing they tacitly agreed with the discovery of the blackbody itself. Alfvén in particular had once been a hands-on electronics engineer and had a great intuition about things space. If he had accepted the blackbody in his gut, that was a great assurance. The Nobel Prizes were awarded.

The discovery of the Penzias-Wilson blackbody was never ever properly evaluated at the very first level. A simple engineering evaluation by everyday engineers in the field was all that was needed. That was the only thing that seems not to have been done - ever. By quickly shifting the matter to the arena of astrophysics, this evaluation was avoided.

The Magus had succeeded in pulling a white elephant out of his hat, none the wiser. The audience broke out in wild applause. The Nobel-givers were a part of this audience.

By the time the Smoot-Mather discovery was placed before the Nobel Committee, all the paeans had been sung and all the bloviating had been done. There were two scientific epics composed that had taken the world by storm. And the thrill of it all! The heroes of these epics were Da Vinci, Magellan and Indiana Jones – all rolled in one. If George Smoot was toiling in the belly of a U-2 spy plane, John Mather was going where no man had ever gone before. If Smoot was looking God straight in the face, John Mather was hearing intimations of immortality. And these are not things I am making up or over-dramatizing.

The paperwork submitted to the Nobel-givers was all shipshape. There had been extensive vetting of the discoveries. And for any die-hard foot-draggers among the Nobel evaluators, Mather's discovery had been followed in its wake by Herbert Gush's discovery. A picture-perfect discovery had been confirmed entirely independently by another picture-perfect discovery.

Let us not forget a very important member of the Dramatis Personae, Stephen Hawking. His carefully composed sound bite not only reverberated around the world, not only helped sell the two aforementioned epics, but was actually cited by the Nobel Prize Committee as a justification for the Nobel Prize.

Awards and accolades were flowing to Smoot and Mather long before they would travel to Stockholm. These also created a compulsion for the Nobel Prize to come forward and fulfill its destiny.

What happened here? In one word, magic. Everybody kept their eyes on the pretty assistants of the Magus while he cut a woman in half.

The main theatrical attribute about the discovery of the acceleration of the expansion of the Universe was chosen to be its shock-jock effect. The world was being trained to hear that the great discoverers chasing exploding stars would report that the expansion of the Universe was decelerating, losing steam that is. In other words, some ho-hum confirmation of Big Bang was in the offing, as if more confirmation was needed. Instead, the world woke up one morning and heard that the Universe was falling apart.

The media mill now went into action, creating the atmospherics and extracting the most mileage out of the shock-jock aspect.

And this aspect was added to when Stephen Hawking, who long rejected this discovery, came round to accepting it around about 1999. As we know from the Smoot-Matter Nobel records, a sound bite from Hawking carries great weight with the Nobel evaluators. So it may be that this stylistic, choreographed change of stance on the part of Hawking was the last piece of assurance the evaluators needed before they would unanimously say: Detta är bra!

The scam here was most intricate and very well-hidden. One could not examine the discovery papers themselves and uncover the scam. One needed the knowledge of the entire scientific and social history leading up to this. One needed the detailed knowledge of the other Big Bang discoveries. One needed constantly to ask: Where is the Perlmutter trap?

Saul Perlmutter, more than any other Big Bang cosmologists, symbolizes the spirit of the Magus. Exploding supernovae, stretching space, burning candles out there – not even Dreamworks SKG could dream up such theatrics.

V-11.4 The great intellectual heist

So the persuasion of the Swedish Nobel-givers by the Big Bang establishment occurred in plain view of the world. The world in fact was made a part of this showy party. The punch bowl was strongly spiked and thrown wide open. When everybody drinks, no one reeks of alcohol.

Add to this the possible desire of the Nobel-givers to move with the times, to stop taking a pedantic view of physics. A lot was going on in the firmament, and the long-traditioned prize did not want to miss out by being too staid and tradition-bound. They needed to venture out and be a part of the firmament history. To this end, they even pressed the poet Robert Frost into service.

So what happened? I think some type of mass hypnosis happened. Nobody was immune, not even the Nobel-givers. How many times has history shown us that mass hypnosis can wreak long-term havoc upon the world? This Big Bang business at least did not cost any lives. All that happened was that the course of the scientific civilization was silently hijacked.

> The old order changeth, yielding place to new,
> And God fulfils himself in many ways,
> Lest one good custom should corrupt the world.
>
> Alfred Tennyson

CHAPTER V-12
BICEP2 Collaboration

V-12.1 The project should never have started

The BICEP2 team set out to find B-mode polarization swirls in the sky predicted from inflation theory. The first scientific assessment that needed to take place was whether or not the prediction was sufficiently defined to merit verification. It seems that it was determined to be so. The prediction gave quantitative values of the strength of the polarized component of the relic radiation. It predicted that the B-mode polarization would have a strength of about 10% of the E-mode polarization, and the E-mode polarization would be about a million times weaker than the intensity of the relic blackbody.

From these the experimenters figured out that detection of B-mode polarization would require a sensitivity of measurement of 1 part in 30 million.

This sensitivity was not achievable by the BICEP2 telescope, no matter what type of instrumentation was put in it. So the question is: Why did this project get started in the first place? How did the project get past the proposal reviewers? Who was minding the store as far as public funds are concerned?

But let us also consider what happened after it did get started. When it was decided to go from the focal plane horn antennas to the "printed circuit" slot antennas, why was not proper expertise sought out? Why were not polarization characteristics of the individual slot antenna measured and compared with that of the dual polarized horn antenna that was to be made obsolete? Why did not the printed circuit designed follow the well-known principles of antenna array design rather than packing the antennas like sardines? Who said that closer the antennas are, the sharper the definition in the sky? Why was the quality of detectors (of secondary importance) bandied around all over the place, and why was nothing at all discussed about the properties antennas (of supreme importance)?

V-12.1 Two views on BICEP2

We will now discuss two views on BICEP2 project as it played out to the end. The first is the view of the scientific establishment, normally the only view that matters and that the world pays attention to. But then for anyone interested, I will also present my view.

The initial reaction of the establishment was one of ecstatic euphoria. They just could not find enough laudation in the English language. Pundits were bloviating all over the place.

But even as this was happening, there were a few rare voices of caution (even before the foreground/background debate started.) For example, Lisa Randall of Harvard University voiced caution publicly (and perhaps courageously), her reserve stemming from, among other issues, the fact that the BICEP2 B-mode signals were far larger than predicted by theory.

One other person merits mention: Edward Witten of Princeton University. It was logical for people to seek out his comment on the discovery, and also logical for Witten to issue a comment. Such a comment would have carried tremendous weight at that juncture. But to my surprise Witten remained silent.

When the foreground/background issue arose, the establishment went quiescent for a while. But as the direction the matter was taking became apparent, there arose a cacophony of homilies whose clear intent was to minimize the damage to the BICEP2 team's reputation and even to put a positive spin on things. This movement became so bizarre that hardcore theoreticians started expressing opinions to the effect that the team had made a great contribution by developing excellent instrumentation. Also, when the Planck satellite data showing that the B-mode signal could be due entirely to galactic dust foreground finally put the discovery to rest, the door was still left open: A little bit of that B-mode signal might have been due to gravitational waves after all! Comments like "The jury is still out" kept popping up. They were not going to give up on a good thing so easily.

In the end it was said that the BICEP2 team made the first observation of B-mode polarization in the sky – a great achievement in itself. After this, could the discovery of gravitational wave be far behind? The team was encouraged to carry on with the good work, undaunted.

There was exactly zero accountability for what transpired right before the eyes of the world. No blame was ever assigned.

Now for my view. As I have said, this project should never have begun. The scientific establishment is to blame for letting it proceed. Having begun the project, it was badly botched – because of ineptness of the researchers, combined with the lack of adequate establishment scrutiny. For this the establishment is to blame. When I exposed the inner workings very fully, the scientific establishment – which had the complete ability to understand my position very completely – chose to look the other way and to party on. The establishment is to blame.

There needed to take place an inquest just like the ones that took place with regard to the reported discoveries of Jan Hendrik Schön and Victor Ninov. Instead, the wagons were circled.

CHAPTER V-13
Planck Collaboration

V-13.1 Planck satellite and the Big Bang Blackbody Spectrum

Planck Collaboration is the name of the scientific team behind the Planck satellite, belonging to the European Space Agency (ESA). The leader of ESA is its Director General, Jean-Jacques Dordain. The leader of the Planck team is Jan Tauber. The European-funded and European-engineered satellite operates at nine discrete frequencies that were carefully chosen to hug the COBE satellite 2.7 K blackbody spectrum.

wikipedia.org   esa

Jean-Jacques Dordain and Jan Tauber

The first results from the Planck satellite were released in 2013. One of the very first calculations they would have made is the CMB absolute intensity in the sky at each of the nine frequencies. In other words, the very first result they obtained – the result that would underpin all subsequent results – was the CMB radiation spectrum. The nine Planck intensity points were expected by the Big Bang establishment to fall smack on the COBE satellite blackbody spectrum.

Planck Collaboration would release a great deal of data and scientific results after that. However, they never ever released that spectrum which was to confirm the COBE spectrum. The reason is most obvious: Planck did not confirm the COBE spectrum. On the contrary, it solidly disproved the COBE spectrum and thus falsified Big Bang theory in its entirety.

Nor did they release the skymaps in total intensity (as distinct from the false "Big Bang-temperaturized" intensity.)

It may be that buried somewhere in the data files they have released there is information to construct the missing spectrum and the skymaps. That would place Planck Collaboration technically in the clear from the charge of suppressing negative data. But why would they engage in such a subterfuge?

Having suppressed the all-important spectrum, Planck scientists then proceeded as though Big Bang were fully intact, and made incremental refinements of Big Bang numbers from the litany of hackneyed Big Bang calculations (age of the Universe, Hubble constant etc.) There was not a stitch of new thought, fresh idea or bold imagination displayed – to match the fine instrument the Planck engineers had delivered. What was on display was pure scientific mediocrity.

In sum, Planck Collaboration started to scam in order to cover up the COBE satellite fraud in particular and to protect Big Bang cosmology in general. They squandered away the great opportunity to bring about a new era of cosmology as an honest science, by merely releasing the CMB spectrum and the dual skymaps of the total intensity and the differential intensity.

V-13.2 Planck Collaboration and BICEP2 Collaboration

When the BICEP2 matter came along, everyone looked to Planck Collaboration for support. There was the issue of how much of the BICEP2 B-mode signal was due to the deep background CMB and how much of it was due to the foreground galactic dust. For the BICEP2 discovery to stand, it was necessary to demonstrate that the latter contribution was minor. And only Planck Collaboration had the data bearing on the foreground emission.

This was the issue as it was presented to the public by both Planck and BICEP2 Collaborations. But it was not the real issue. The real issue was that the BICEP2 B-mode signals were the result of total instrumental botch-up. The B-mode sky swirls were artifactual.

Planck Collaboration, a part of ESA, had access to the most advanced engineering knowhow in the field. Therefore they, if anyone, would know that BICEP2 was an instrumental botch-up. Therefore, they would not engage in such a collaboration with BICEP2 that by the act itself would aver that the BICEP2 measurements were scientifically sound.

But they did. They showed that the foreground was the dominant contribution. This made the BICEP2 discovery of the primordial gravitational waves go away, but solidly confirmed that BICEP2 was the first instrument to observe B-mode swirls in the sky. Planck Collaboration knowingly covered up the BICEP2 instrumental botch-up.

It should be mentioned that Planck Collaboration has many American members, and BICEP2 Collaboration has European members. So the line of demarcation is not all that clear.

Planck satellite has acquired very high quality, highly valuable data. There is no question about this. But as long as this treasure trove is considered the private playground of Big Bang cosmologists, everyone else would feel excluded, not welcome. The treasure trove would remain de facto a private property. The lasting legacy of Planck satellite will be that they repeatedly covered up American botch-ups.

CHAPTER V-14
Beware of geeks bearing predictions

Sometime in 2008 the world had a rude awakening. We learned about how abstract Wall Street investment instruments like Collateralized Debt Obligations and Credit Default Swaps threatened the economy. Some of these were actually designed by physicists and mathematicians who saw a new fertile field in the financial arena to apply their skills. The renowned investor Warren Buffett wrote to his shareholders:

Constructed by a nerdy-sounding priesthood using esoteric terms such as beta, gamma, sigma and the like, these models tend to look impressive. Too often, though, investors forget to examine the assumptions behind the symbols. Our advice: Beware of geeks bearing formulas.

Our discussion thus far has been about experimenters. It is easy to fall into the trap of thinking that the Big Bang theorists are blameless. But they are very much a party to the grand deception. So let us talk about the priesthood in cosmology.

Now, if I were a theoretician and had made some theoretical studies to predict that the star Alpha Centauri is made of antimatter, I would have a good gut feeling about how well-formed this prediction is. Is this an idea that can be handed over right away to the experimenters? Is this an idea ready for the investment of tens of millions of dollars of public funds? My conclusion would be that this is a very long shot, not worth spending great funds on. I would not want to be responsible for such a commitment. If some hotshot young bucks from big name universities came to me and said they wanted to make experimental verification of my prediction and that they would have no problem getting funds, I would gently dissuade them.

That is what a theoretician's responsibility is: To know in his physics gut what he has produced and not to oversell it and not to sit passively as funds are solicited by others for this purpose.

The Big Bang theoreticians likewise make theories and predictions. But when men like Alan Guth and Andrei Linde encourage – actively or passively – the lavishing of public money to verify their predictions, they are being irresponsible. For they have to know full well that their fanciful omphaloskepsis does not merit this expenditure. Either that, or they have no physics in their gut at all.

To make the above point more concrete, let me cite what Andrei Linde – in a recent unguarded confessional moment – said about his idea that the BICEP2 team was trying to verify:

I always live with this feeling: What if I believe in this just because it is beautiful?

And yet he condones the spending of tens of millions of taxpayer dollars on his private fantasy.

Suppose in the example about Alpha Centauri, some people went ahead anyway and collected satellite evidence that I am right. Would I exalt over it? No. I would have grave doubts even though it is my own theory that has reportedly been verified.

But the inflation theorists gloried (in the noble way they glory) when they heard about the BICEP2 confirmation of their ideas. Again, Andrei Linde's comment here is revealing:

If this is true, this is a moment of understanding of nature of such a magnitude that it just overwhelms,...

We have here scientists that have been prematurely put on such lofty pedestals that they could not developed a sense of an intellectual's responsibility vis-à-vis the society down below.

CHAPTER V-15
The Vatican

V-15.1 The Cosmic Egg and the Vatican

A link between Big Bang cosmology and the Catholic Church has existed from the moment the scientific idea was born in late 1920s and early 1930s. The idea was not known by the name Big Bang then, but went under various shorthand descriptions such as the Primitive Atom and the Cosmic Egg. There are two reasons for this nexus.

First, the progenitor of the idea, Georges Henri Joseph Édouard Lemaître, was a Roman Catholic priest who was also a professor of physics at the Catholic University of Louvain in Belgium. So it is natural to think that the ideation must have occurred within a specific religious and spiritual worldview. A priest after all is expected to be fully imbued with this worldview and everything else he thinks or does is expected to flow from there. Lemaître was in evidence in his scientific community wearing his priestly attire, and that imagery surely had an impact.

Second, Lemaître's central idea of creation out of nothing resonated well with the Church. Here was scientific support for theology. The then leader of the Catholic Church, Pope Pius XII, praised Lemaître for finding scientific support for the biblical origin.

However, there are conflicting reports as to whether Lemaître saw any connection between the Primitive Atom origin and the biblical origin. He is on record as saying that his scientific theory has no more to do with religion than the Binomial Theorem has to do with religion. He reportedly convinced the Pope not to promote the religious nexus.

V-15.2 Science and the Vatican

The Vatican has a long history of association with science. This association is formalized through their Pontifical Academy of Sciences established in 1936. This institution descends from a line of ancestral bodies the first of which had Galileo Galilei as its President. At any given time, the Pontifical Academy inducts into its roster prominent contemporary scientists in a wide variety of fields, and from all over the world. Lemaître was the President of the Academy in the 1960s.

Thus it can be reasonably expected that the Academy has available to it the highest quality scientific advice and guidance.

One out of six human beings of this planet is a Catholic. So a billion people follow the lead of the Vatican. This is why what the Vatican says or does in science should be of interest to all of us in general, and to scientists especially.

It is certainly a good thing for religious faiths to stay in close contact with science. It is probably a good thing for scientists to stay in touch with some religious faith – in a conscience and spirit sense. So I wholeheartedly admire Vatican's general interest in science.

Galileo Galilei was a long time ago – in a different time, in a different world. Science recovered from this dark chapter and triumphed. Galileo's legacy is today one of the most magnificent ones in the civilization's lore. Today's Vatican – as far as I am concerned – has no blame for what happened four hundred years ago. And in any case, they just forgave Galileo.

V-15.3 Big Bang and the Vatican

But here is the thing: I have to resurrect Galileo now because of the uncanny parallels (or more correctly, anti-parallels) between the Catholic Church's stance on Galileo Galilei then and on John Cromwell Mather today. If the Church ostracized Galileo for his correct view of the firmament then, they are blessing Mather for his false firmament. If this view were allowed to proceed unchecked, then four hundred years from today children will read a false firmament history – the Vatican firmament history. It will extol a trio of great Genesis Scientists: Copernicus-Galileo-Mather.

What is it with the Vatican that causes the above streak to surface again? Why are they trying proactively to influence the course of the development of the scientific view of the Universe?

V-15.4 The patronage

There are at least four powerful and prominent forums through which the Catholic Church can express/exert itself in matters of science:

The Catholic University of Louvain, Belgium
The Catholic University of Notre Dame, USA
The Pontifical Academy of Sciences, The Vatican
The Rimini Meetings, Italy

As I recounted before, Big Bang cosmology was conceived in Louvain – in the cloisters of a Church there. Try to imagine that serene surrounding and the kinds of thoughts it might inspire and incubate in a most intelligent mind given unconditionally to God.

In 2004, emboldened by the Smoot-Mather 'discovery', the Catholic Church boldly published its official position:

... the Universe erupted 15 billion years ago in an explosion called the 'Big Bang' and has been expanding and cooling ever since.
...

THREE POPES ON BIG BANG

Pope Pius XII, Pope Benedict XVI, and Pope Francis

If we look back into the past at the time required for this process of the "expanding universe," it follows that, from one to ten billion years ago, the matter of the spiral nebulae was compressed into a relatively restricted space, at the time the cosmic processes had their beginning.

Pope Pius XII (November 1951)

According to the widely accepted scientific account, the universe erupted 15 billion years ago in an explosion called the 'Big Bang' and has been expanding and cooling ever since. Later there gradually emerged the conditions necessary for the formation of atoms, still later the condensation of galaxies and stars, and about 10 billion years later the formation of planets.

Pope Benedict XVI (endorsing the position statement as Cardinal Ratzinger in 2004)

The Big Bang, which today we hold to be the origin of the world, does not contradict the intervention of the divine creator but, rather, requires it.

Pope Francis (October 2014)

At least three Popes commented on Big Bang cosmology.

After the Smoot-Mather Nobel Prizes were awarded, the other three organizations mentioned are especially in evidence as plugging it proudly and loudly, the self-evident science fraud notwithstanding. His Holiness Pope Benedict XVI personally greeted George Smoot.

In April 2009 the University of Notre Dame invited John Mather to deliver their distinguished John A. Lynch Lecture and feted his discovery.

The Rimini Meetings happen under the auspices of the Vatican. It was founded by a priest there. In the Rimini Meetings 2009 held in August, John Mather was feted prominently before a huge audience. Mather's face was projected on a floor-to-ceiling video screen – and a photograph of this was later bandied about.

In the spring of 2011 the University of Notre Dame conferred on John Mather the honorary degree of Doctor of Science. On 22 May 2011 Mather stood in a commencement gown emblazoned in gold, alongside such powerful figures as the US Defense Secretary Robert Gates. The Iranian Peace Nobel Laureate Shirin Ebadi was also invited to this stage, but she was absent. In the official photograph of the gowned luminaries Mather and the former NASA Administrator Michael Griffin were placed next to each other, both smiling broadly. It was as though the University (the Church?) were creating a public image that there had occurred no rift between the two as I have conjectured (Appendix A).

In composing the *Doctor of Science – Honoris Causa* citation for John Mather, Notre Dame engaged in very crafty language-smithing to avoid mentioning Mather's actual official Nobel discovery (the blackbody spectrum form of the Cosmic Microwave Radiation) which was fraudulent. Instead, here is how the Notre Dame citation went:

(Mather) transformed the study of the early Universe from a largely theoretical pursuit into direct observation and measurement...

Technically speaking, this citation does echo what the Nobel web page said in the general terms about the COBE satellite (of which Mather was the overall scientific manager.) This is even more evidence of trickery. Notre Dame found some language in the Nobel web page that they could safely substitute for the official discovery citation – which has now become problematic. In other words, they will engage in trickery for their purpose rather than simply not give Mather the honor – which is what most institutions without any ulterior agenda would do.

An organization cannot distance itself from what its integral members do – especially when such doings are known to and endorsed by the leaders, especially when these go on for years, and especially when questions about their conduct have been raised for years.

Therefore the Catholic Church is in evidence today as pursuing a consistent system-wide policy of supporting and promoting Big Bang.

V-15.5 The New Crusade?

Why would the Catholic Church develop this elaborate contact with Big Bang cosmology? To find the answer to this we have to broaden our perspective.

Religions of the world have to grow, or at least constantly replenish their adherents, in order to survive. The leaders of the Vatican are deep intellectuals who take time to think of long range prospects for Catholicism. In the cyber civilization of the twenty-first century, it makes sense for a great religion to develop scientific underpinnings. This would be the appeal to the emasses. It makes for a safe, secure, peaceable theology on

351

high ground, devoid of aspects on which clashes or conflicts with other religions can occur.

All this is good for the Vatican, and non-Catholics should have no reasons to object. Nor should science have any reasons to object.

So what is the problem with the Vatican patronage of Big Bang cosmology? None whatsoever if this science were conducted as science. The problem is that Big Bang is a fraud and the Vatican is supporting this fraud to the hilt. And in most subtle ways, this support can ensure the prospering of the venture in the *scientific* arena.

Thus in my view, Big Bang cosmology is much larger matter than it appears on the surface. It is a 'religious' hijacking of the scientific civilization. It heralds the decline and fall of this civilization.

Of course there is an alternative interpretation I am open to: A succession of Popes has been receiving receiving bad scientific advice from the Pontifical Academy of Sciences.

> ...the existence of this Pontifical Academy of Sciences, of which in its ancient ancestry Galileo was a member and of which today eminent scientists are members, without any form of ethnic or religious discrimination, is a visible sign, raised amongst the peoples of the world, of the profound harmony that can exist between the truths of science and the truths of faith....The Church of Rome together with all the Churches spread throughout the world, attributes a great importance to the function of the Pontifical Academy of Sciences. The title of 'Pontifical' given to the Academy means, as you know, the interest and the commitment of the Church, in different forms from the ancient patronage, but no less profound and effective in character...John Paul II (1979)

The central role of the Pontifical Academy of Sciences in shaping the Vatican scientific worldview.

APPENDICES

Appendix A
The spring 2008 dust-up at NASA Headquarters

Following the announcement of the Nobel Prize to John Mather on 2 October 2006, his employer NASA rose to the occasion. Their very own, home-grown Nobel Laureate – the first Civil Service Nobel Laureate in fact – needed to be recognized and anointed properly within the Agency. On 2 April 2007, two major announcements were made. First, Alan Stern – a career space scientist who worked in the private industry – was appointed NASA's Associated Administrator. He was responsible for the Science Mission Directorate (SMD). The scientific establishment applauded this appointment. Stern then announced the creation of a new (or long unfilled) position, Chief Scientist of the SMD, and installed John Mather in it.

In so appointing Mather, Stern described him thus:

John Mather is a scientist of legendary reputation, technical ability and space science mission experience. His office will provide independent scientific advice to me to guide decision making regarding all aspects of the NASA science program.

John Mather himself wrote thus of this appointment in his Nobel autobiography:

I am also broadening my perspective one more time, trying to learn about the entire range of space science, and helping to guide NASA science towards the discoveries of the future. On April 2, 2007, I will take on the job of Chief Scientist of the Science Mission Directorate of NASA, so I will have the opportunity and responsibility to advise NASA on the proper balance of scientific programs from Earth science to cosmology. The panorama of amazing research programs is almost overwhelming, and I am looking forward to seeing it.

The Stern-Mather duo, with full blessing and backing of the then NASA Administrator Michael Griffin, now represented the beginning of a new era for the SMD. Mather's lecture circuit calendar was very full.

Now pause a moment and appreciate the depth and the breadth of the discovery Mather made. Stephen Hawking was not exaggerating when he described the Smoot-Mather discovery as

...discovery of the century, if not all time.

Nor was Per Carlson, then Chairman of the Nobel Committee for Physics, when he said:

It is one of the greatest discoveries of the century. I would call it the greatest.

And this pertains not only to science. What Mather did would in effect extend the Book of Genesis all the way back to the original Origin. All other discoveries man has made pale in comparison.

On 5 April 2007, I published a web page on the Internet, alleging that Mather's discovery was fraudulent. I also explained in my site why I resorted to Internet publication in my own website rather than the submitting a paper in a refereed journal of the physics establishment.

I did not expect that there would be any impact of my site on the scientific establishment. I published and subsequently publicized my site to fulfil some need of my own. As time went by without any response to my site, I increasingly heightened the rhetoric and cast my net wider. I made sure to let NASA HQ and NASA OIG (Office of the Inspector General) know about my allegation through their official email addresses.

Nothing happened. John Mather continued to be feted by the world.

On 26 March 2008, less than a year after my site was posted and less than a year after Stern's coming to NASA, the planetary and space science establishment woke up to hear that he had resigned from NASA! There arose a cacophony of great sadness and great lamentation from within the scientific establishment. A true friend of science was being booted out by a bureaucratic NASA.

There was speculation that Stern had a falling out with his boss Michael Griffin. The two apparently had very different philosophies about which programs to prioritize, budget-wise speaking. Stern would later confirm this speculation. From Griffin's office, however, word was put out that Stern was doing things without telling Griffin. This would seem to suggest that the "resignation" was not 100% voluntary. Also, it was very sudden, and Stern did not have any comparable positions lined up. He simply went "back to the pavilion" – as they say in cricket.

So what really was the reason behind Stern's summary departure from the top executive post in NASA, with a shot at the No. 1 position? Would any careerist jeopardize this situation over something as intangible as philosophy? There is room for great disagreements in such quarters, and people do not have to leave because of them. And surely Griffin knew all about Stern's philosophy when he hired Stern less than a year ago. Stern was a very well-known and active player in the NASA community.

Alan Stern reportedly wanted to promote the cause of basic science from within NASA. He found himself in the appropriate power position to do so. But apparently, Griffin turned out to be an obstacle. In the spring of 2008 an astute careerist might have figured that a Democratic administration would soon come to

Washington DC and Griffin would be gone from NASA. The careerist would then have decided to lie low for just a bit, and not pick any precipitous fights with Griffin. Within a year, the obstacle would be gone and he would have a clear shot at what he wanted to accomplish.

In a NEW YORK TIMES article dated 1 January 2008, Stern is reported as saying, about scientists who were pressing him about funding issues and whom Stern had earlier rebuffed, the following:

"When they came to believe I was serious and had my boss's backing," Dr. Stern said, "they took it seriously. They quickly found a way to erase that bill."

So it would seem that even as late as January of 2008, Stern had Griffin's backing on whatever funding philosophy Stern was pursuing.

No matter how we look at it, the cover story put out in the media does not add up.

Now, here is something even stranger: The other event of interest here. On the same day that Stern resigned, or within a day, John Mather resigned the Office of the Chief Scientist, SMD. His colleague Robert Park at the University of Maryland reported in a blog that Mather's resignation was not connected to the Stern resignation. Even more strange, in the face of the shocking news, Mather told Park that "his resignation was not at all abrupt." Having resigned, Mather then declared on his own that he will continue to spend time at the SMD "until further notice."

Immediately following this, NASA issued a Press Release stating that Mather was being sent back to the Goddard Space Flight Center (GSFC), Greenbelt, MD, and assigned to work full-time on the James Web Space Telescope (JWST). It was not explained why he had to vacate the Office of the Chief Scientist in order to do this. Both GSFC and NASA HQ are within Mather's commute distance. Why could he not have kept the HQ post while attending to any pressing JWST matters? The Chief Scientist, after all, was meant to be a person who defined his own job. He was not a person who was to be assigned tasks. Anyway, the "further notice" came when this high visionary office was quickly filled with a non-stellar bureaucrat from within the NASA ranks – perhaps even the old boy network.

I have myself worked on LANDSAT and INTELSAT projects, and I have a pretty good idea about who does what and when and at what stage in such large and complex programs. The above explanation for Mather's departure from NASA Headquarters does not hold water for me.

According to a Wikipedia entry, the status of JWST at this point in time was as follows:

In March 2008, the project successfully completed its Preliminary Design Review (PDR). In April 2008, the project passed the Non-Advocate Review.

So all the scientific design work had actually been completed. There was no conceivable reason for pulling a NASA HQ science leader off his job and sending him back to GSFC to work full-time on the JWST. Even if there were some unexpected fire to put out with the JWST, it did not necessitate Mather's resignation. This cover story was completely transparent to me, and in my view, meant to be transparent. It was meant to convey a message to the scientific establishment, a message that for a number of reasons could not be spelled out at that juncture.

I think it is fair to conclude that something is out of kilter here. Why had the great American superhero with a golden *S* emblazoned across his chest come to this place, without any protestations or comments from his powerful constituency, the academic physics establishment? Did Michael Griffin have enough clout to render a great American hero a nonentity, for no clear reason at all? Definitely not.

Can two precipitous resignations at the highest level of the Agency on the same day be unconnected? Why is it that neither cover story makes sense? Why is that subsequently, neither of the two brilliant shooting-star careers recovered? What could be a reason that would make NASA Management feel completely secure that they could defend these changes to anyone? Certainly, it was quite probable that they might have had to explain things to the NASA Oversight Committee of the US Congress and to President George W. Bush (Griffin's boss). And why is it that no public inquiries were ever held by this Committee on this deeply embarrassing Government spectacle?

My own interpretation was that the Nobel Hero could be taken down in this way only if there was something the matter with the Nobel Prize itself. Michael Griffin must have commissioned a confidential engineering study on the COBE satellite, and the results of that study must have been devastating. Griffin felt he needed to do *something* on record so that the US Government would not be on record as knowingly covering up bad business. It may be that a well-meaning but misadvised Stern sided strongly with his friend Mather, and that is how the whole situation came about.

Appendix B
NOVA program *Hunting the Edge of Space*

A television program on the history of cosmological observations titled *Hunting the Edge of Space* was released in April of 2010 within the Public Broadcasting Service (PBS) NOVA program. It covers the great telescopes from the days of Edwin Hubble to the present day.

The relevant portion of the program for us is Hour 2, *The Ever-expanding Universe: How telescopes have expanded our view of the Universe.* It brings the reader up to the discovery of the Big Bang relic radiation by Arno Penzias and Robert Wilson, and then abruptly jumps to the Wilkinson Microwave Anisotropy Probe (WMAP) Satellite, which was the successor to the COBE satellite. This latter satellite – by far the most noteworthy of all the Big Bang satellite – has been surgically expunged from the history of cosmological discoveries.

In this program there appear many well-known members of the scientific establishment, including some from NASA. Some of them – like Wendy Freedman and Matt Mountain – are quintessential Big Bang insiders.

For the purpose of public education I reproduce a short portion of the transcript below:

NARRATOR: *Penzias and Wilson have unwittingly found the first physical evidence of the Big Bang. Their radio telescope has picked up the afterglow of the beginning of the Universe. It remains one of the most important discoveries of all time.*

ARNO PENZIAS: *I had no idea that we were listening to the echo of creation.*

NARRATOR: *In 1978, Penzias and Wilson received the Nobel Prize.*

ARNO PENZIAS: *To be on a list with Albert Einstein, to be on* *that same roll, was almost too much, too much to bear. I just couldn't think of comparing myself against the people who have won Nobel Prizes.*

WENDY FREEDMAN: *In 1964, when the background radiation from the Big Bang was discovered, it was, for the first time, direct evidence that there was a hot Big Bang, an origin to the Universe.*

NARRATOR: *Penzias and Wilson have found direct proof for the Big Bang, but it will take another 37 years and far more sophisticated microwave technology, before we discover how the Big Bang formed the Universe we see today.*

[????]

Heading to an orbit, 1,000,000 miles from Earth, is the Wilkinson Microwave Anisotropy Probe, or WMAP, a super-advanced version of Penzias and Wilson's giant antenna, armed with two reflecting telescopes.

WMAP's mission is to examine the afterglow of the Big Bang in extreme detail and to try and find out why galaxies formed.

CHARLES BENNETT *(Johns Hopkins University):* *With WMAP we were trying to look way back to the very, very earliest times in the Universe.*

NARRATOR: *After a year of recording, the first results are mapped.*

WENDY FREEDMAN: *The WMAP observations were incredible. Instead of a smooth background radiation, you could measure to 1,000th-of-one-percent changes in the temperature, across the sky.*

NARRATOR: *WMAP shows that, actually, there are tiny fluctuations in temperature.*

The place where I have added the question marks is the

place that belonged to the pioneering COBE satellite, the mother of all Big Bang satellites. I cannot say if this history was added and later purged, or never added.

But it certainly is most baffling. It is like producing a PBS-NOVE program on the history of the DNA molecule, and surgically expunging any references to Francis Crick and James Watson.

The strangest thing is that the scientific community, and especially the Big Bang establishment, did not react to this at all. The Big Bang insiders who participated in this program simply went along with this.

Appendix C
Referrals of the science fraud allegation

The Referrals

Following the posting on the Internet of my exposé on the COBE satellite on 5 April 2007, I sent a number of referrals to pertinent organizations/people. These were sent to the published official email addresses or through published official webmail channels. By referral I mean bringing my web page to the addressees' attention. This is all that was within my ability to do.

It should be noted that these referrals were one-sided in the sense that I initiated an act that remained unacknowledged. Except for some electronic receipts for web mail communication, I never received any responses to these referrals. I have no knowledge of what transpired at the other end. I fully understand and appreciate that one cannot involve someone in something by simply sending an email to that someone.

This is the present link to the subject Internet post:

http://www.bibhasde.com/blackbody.html

Why did I not submit a critique to a refereed scientific journal instead?

The usual response of the scientific establishment to any criticisms raised by an outsider is to say: "Submit your criticism to a refereed journal."

I have been sent down that path many times before in other contexts. When you try to publish thusly, they have essentially moved you to an out-of-view private arena where they function exclusively on their terms. They can offer a plethora of excuses written anonymously on why they cannot publish. If you pursue this further, they will tie you up in knots. Years of your life will pass by and nothing will happen. Such is the collective and organized resistance in this community against outside dissent. Just think about this: If there is a serious scientific issue, why should it matter how or where it was raised?

The Big Bang dissidents – some of them card-carrying members of the scientific establishment – have also tried the refereed publication route. They have sometimes managed to publish their criticisms. These have been completely ignored.

At any rate, with the present issue we are far beyond the scientific debate phase. The award of the Nobel Prize has closed that phase. We are now dealing with science fraud and the proper arena to air this is the public arena.

Some may see in the following a somewhat laughable effort on my part of tilting at windmills. It certainly seems so to me as I look back. At the same time, it is also true that something in these quixotic forays triggered something that caused the United States to confine her very first Civil Service Nobel Laureate to his pre-Nobel Prize job classification. There is no humor there.

The Nobel Foundation

On 14 June 2007, I referred the site to the Nobel Foundation (*comments@nobelprize.org*).

On 21 June 2007, the email (Figure C-1; misspelling noted) was sent to the individual members of the 2006 Physics Nobel Committee, and individual leaders of the Nobel Foundation and the Royal Swedish Academy.

NASA

On 23 April 2007, the web page was brought to the attention of the *Office of the Inspector General, NASA* (Figure C-2).

On 8 May 2007, a link to this site was sent to the NASA Headquarters through their *ASK NASA* link (Figure C-3), with the following question:

"What do you have to say about the detailed allegation of science fraud at NASA at the following public site?

http://www.geocities.com/bibhasde/blackbody.html"

The Congress of the United States

On 26 April 2007, the site was referred to the Committee on Science, U. S. House of Representatives, led by Representative Bart Gordon (Figure C-4).

On 6 February 2009, a web mail was sent to Senator Bill Nelson, Chairman of Senate Oversight Subcommittee on NASA (Figure C-5).

The United States Government

On 18 August 2007, the Office of Science and Technology Policy (OSTP), the ultimate watchdog on research misconduct involving Government employees or Government funds, was informed on the matter by email (their preferred means of communication). See Figure C-6.

A letter was sent by email on 29 December 2008 to the President's Council of Science Advisors (Figure C-7).

On 24 August 2010, I sent a letter to President Barack Obama through this channel about science fraud committed by the US Government scientist John Mather (Figure C-8). The Head of NASA reports directly to President Obama.

On 18 October 2010, John Mather was invited to a White House event to inspire young people to pursue science. Also invited was his boss, NASA Deputy Administrator Lori Garver, who was aware of the allegation. On 21 October 2010 I sent the message in Figure C-9 to the White House.

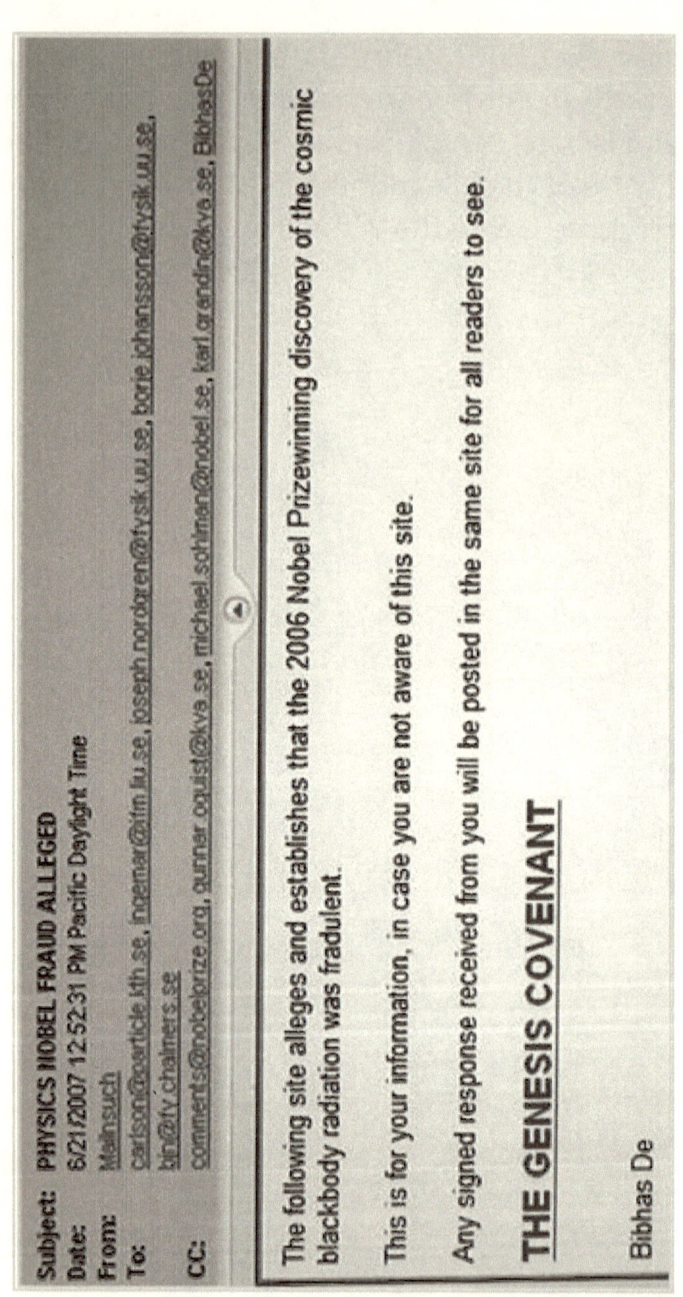

Subject: PHYSICS NOBEL FRAUD ALLEGED
Date: 6/21/2007 12:52:31 PM Pacific Daylight Time
From: Mainsuch
To: carlson@coarticle.kth.se, ingemar@itm.fiu.se, joseph.nordgren@fysik.uu.se, borje.johansson@fysik.uu.se, bjn@fy.chalmers.se
CC: comments@nobelprize.org, gunnar.oquist@kva.se, michael.sohlman@nobel.se, karl.grandin@kva.se, BibhasDe

The following site alleges and establishes that the 2006 Nobel Prizewinning discovery of the cosmic blackbody radiation was fraudulent.

This is for your information, in case you are not aware of this site.

Any signed response received from you will be posted in the same site for all readers to see.

THE GENESIS COVENANT

Bibhas De

Figure C-1: The Physics Nobel Prize Committee was informed on the research misconduct allegation against John C. Mather.

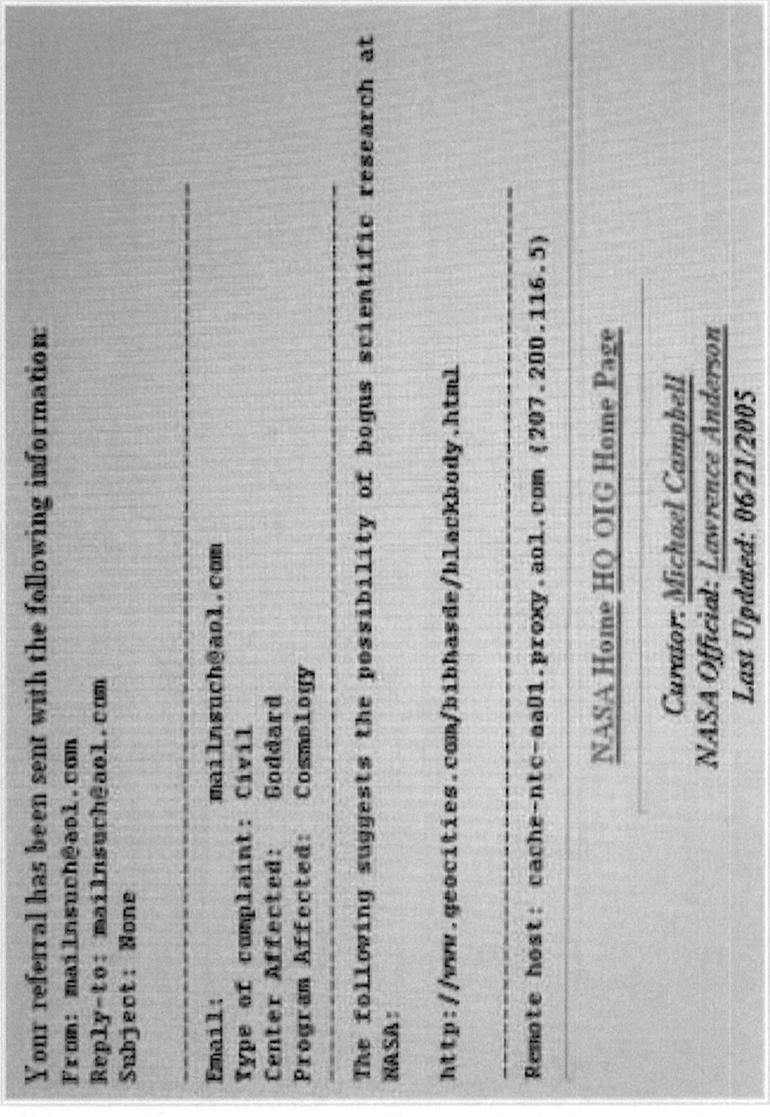

Your referral has been sent with the following information:
From: mailnsuch@aol.com
Reply-to: mailnsuch@aol.com
Subject: None

Email: mailnsuch@aol.com
Type of complaint: Civil
Center Affected: Goddard
Program Affected: Cosmology

The following suggests the possibility of bogus scientific research at NASA:

http://www.geocities.com/bibhasde/blackbody.html

Remote host: cache-ntc-aa01.proxy.aol.com (207.200.116.5)

NASA Home HQ OIG Home Page

Curator: Michael Campbell
NASA Official: Lawrence Anderson
Last Updated: 06/21/2005

Figure C-2: NASA-OIG was informed on the research misconduct allegation against John C. Mather, through their official web link.

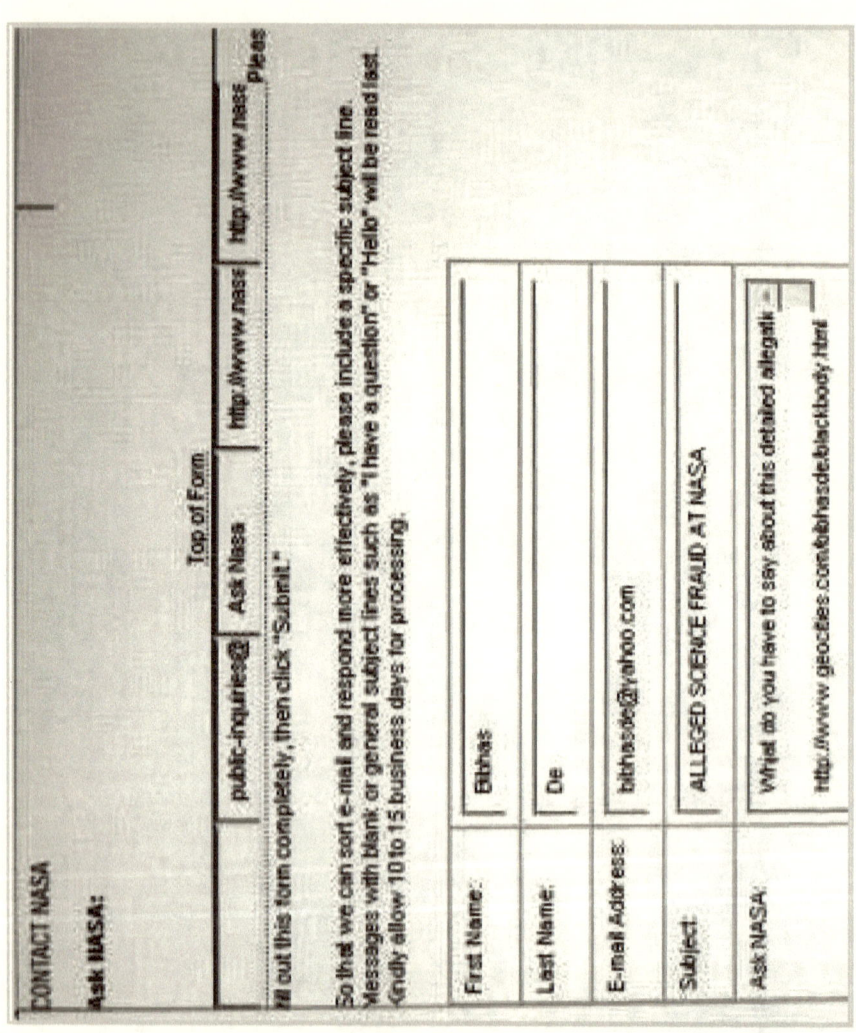

Figure C-3: NASA was informed on the research misconduct allegation against NASA Chief Scientist John C. Mather, through their official web link.

Committee on Science
DEMOCRATIC CAUCUS
U.S. HOUSE OF REPRESENTATIVES   REP. BART GORDON, RANKING MEMBER

Thank you for contacting us!

The following information was received by our office.

04/26/2007 18:51

Bibhas R. De

mailnsuch@aol.com

Can we share your name and information with Member Offices? Yes

[phone]

In the following online article I allege a gigantic science fraud at NASA. This is not a crackpot matter. THE GENESIS COVENANT Is the Big Bang Blackbody Bogus? http://www.geocities.com/bibhasde/blackbody.html Sincerely, Bibhas R. De, Ph. D. Physicist www.bibhasde.com

Figure C-4: The referral to the U.S. House of Representatives, Committee on Science, headed by Rep. Bart Gordon.

CATEGORY: Space
SUBJECT: NASA Oversight/Science Fraud

There is a huge problem with NASA which their leadership is aware of and which they are covering up. It is that their most famous scientist, their homegrown Nobel Laureate is a science fraud. It concerns a fraudulent discovery involving a NASA Satellite, and a fraudulent Nobel Prize.

This is NOT a crackpot allegation.

You can read the now two years old allegation here: http://www.geocities.com/bibhasde/blackbody.html

Thank you for sending me an email. I appreciate receiving your input.

Figure C-5: The referral to Senator Bill Nelson, Chairman of Senate Oversight Subcommittee on NASA.

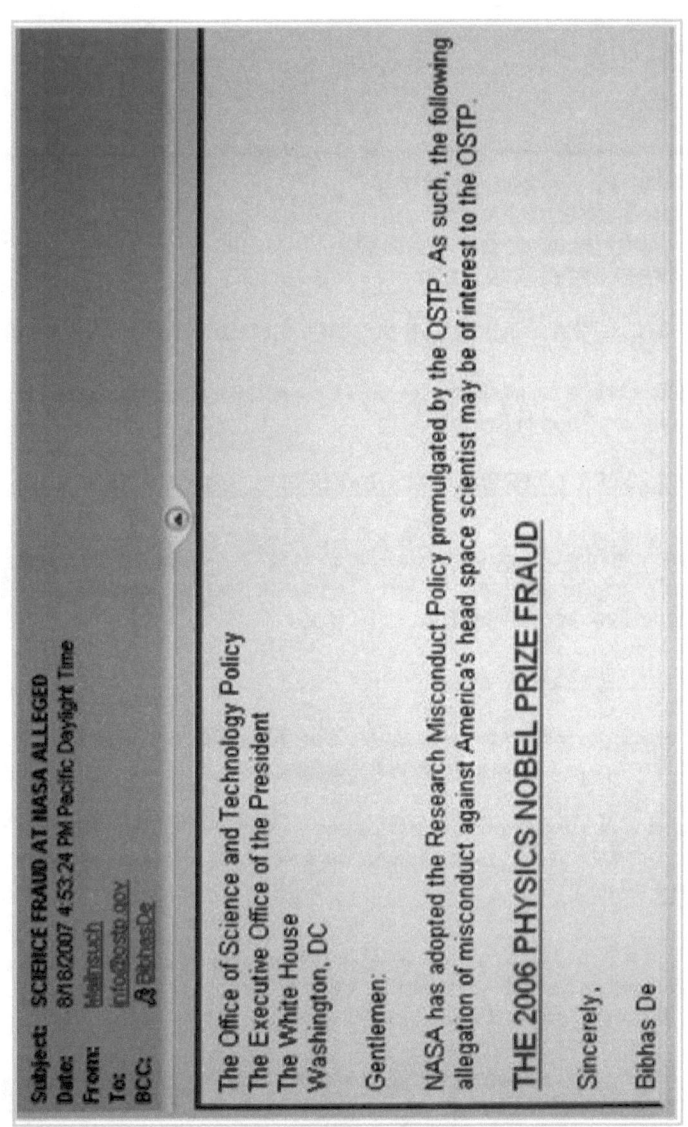

Subject: SCIENCE FRAUD AT NASA ALLEGED
Date: 8/18/2007 4:53:24 PM Pacific Daylight Time
From: bdainsuch
To: info@ostp.gov
BCC: BibhasDe

The Office of Science and Technology Policy
The Executive Office of the President
The White House
Washington, DC

Gentlemen:

NASA has adopted the Research Misconduct Policy promulgated by the OSTP. As such, the following allegation of misconduct against America's head space scientist may be of interest to the OSTP.

THE 2006 PHYSICS NOBEL PRIZE FRAUD

Sincerely,

Bibhas De

Figure C-6: The OSTP was informed on the research misconduct allegation against John C. Mather.

Dr. John Holdren (john_holdren@harvard.edu)
Dr. Eric Lander (lander@broad.mit.edu)
Dr. Jane Lubchenco (lubchenco@oregonstate.edu)
Dr. Harold Varmus (varmus@mskcc.org)

SUBJECT: THE GREAT AMERICAN SCIENCE FRAUD AND ITS COVER UP

First of all, please believe me when I tell you this is not a crackpot matter. This letter is being simultaneously posted on an Internet site:

OPEN LETTER TO PRESIDENT OBAMA COUNCIL OF SCIENCE ADVISORS

For nearly two years now, a gigantic science fraud involving US scientists, US Government agencies and US taxpayers' money has been brought to light. The Swedish Nobel organization has been hoodwinked into anointing this fraud with a Nobel Prize. Please visit:

NASA SCIENCE FRAUD

No one has done anything about this, including the White House Office of Science and Technology Policy (OSTP) whose jurisdiction it was to deal with this matter.

Your colleagues in the scientific establishment have remained completely silent. The worst of them are thus dishonest and dishonorable, and the best of them are choosing to look the other way. Such is how the academia is behaving.

It is possible - but I am not at all sure about this - that the only person who has displayed any honesty in this matter is a much-vilified bureaucrat, NASA Administrator Michael Griffin. He did not have any role in this fraud, but when confronted with it, he may have done something bold about it.

The question now is this: Given all this talk about how wonderful President Obama's science picks have been, what are going to do about *this*? Are you going to serve him well or are you going to fail him right off the bat? And who shall apologize to the World?

Figure C-7: The referral letter to the President's Council of Science Advisors.

24 August 2010 (SUBMITTED VIA WEBMAIL)

President Barack Obama
The White House, Washington, DC

Dear Mr. President:

I will get right to my point: The greatest science fraud in history committed by a US Government Scientist who garnered a Nobel Prize for this in 2006. His employers, NASA, are aware of this allegation – first exposed by me on 5 April 2007. The World learns about this continuously through my Internet campaign: a Let-the-people-know platform to supplant a media that remains silent on this subject.

The fraud has been documented by NASA themselves: Their satellite experiment failed miserably in the sky; on Earth "the greatest discovery of all time" was proclaimed.

This matter was referred unsuccessfully to: Office of Science and Technology Policy; The President's Council of Science Advisors; Congressional Oversight Committees; and of course, NASA and the Nobel Foundation.

The fraud is being erased as I write. This summer NASA had a replica of this Nobel Medal flown in the Space Shuttle Atlantis. Also this summer a symbolic photo of the NASA Deputy Administrator and a senior Astronaut flanking this scientist was released. Other similar events suggest that formidable power groups and interests – enabled by a media that is choosing to look the other way - are arrayed to shield this scientist from accountability.

America has shown great moral outrage in cases of past science fraud. Now the World awaits a clear admission and an unqualified apology from America, and science awaits being made wholesome again.

Respectfully yours,

Bibhas De

Figure C-8: The referral letter to President Barack Obama.

Sent 21 October 2010 3:20 pm through white House Webmail

On 24 August 2010, I respectfully sent a letter to President Obama through this channel about science fraud committed by the US Government scientist John Mather.

On 18 October 2010 the same scientist was invited to a White House Dinner. Also invited was his boss, NASA Deputy Administrator Lori Garver, who is aware of the allegation.

In absence of any other response I have assumed that the President has rejected the allegation.

Bibhas De

Figure C-9: Follow up message to the White House
(see Figure C-8.)

APPENDIX D
Speculation: Spontaneous reddening of starlight

The following rather expansive speculation concerns how the light from an astronomical object can spontaneously shift to lower frequencies during its travel through empty space.

Although the idea stems for me from my view of the photon, it may not be crucially dependent on that view.

LIGHT

Let us discuss the optical region of the spectrum, and for clarity, let us discuss light of a "single" frequency v. If a star has a spectral luminosity L_v, then the intensity of this light at a distance R from the star is I_v:

$$I_v = L_v / 4\pi R^2 \qquad (D\text{-}1)$$

The spectral energy density u_v at this location due to the star is

$$u_v = I_v / c. \qquad (D\text{-}2)$$

PHOTON

The photon in my view has a size (just physical size.) The photon "at rest" is a structure made of magnetic field in empty space, and this structure at a given frequency is defined by an inverse length parameter a. I will now say that the length is typically the wavelength λ ($\sim 1/a \sim c/v$). So – not surprisingly – the photon occupies a volume of the order $V \sim \lambda^3$. By the principle of superposition, the same volume of course can simultaneously be occupied by an unlimited number of photons in a continuum of frequencies.

LIGHT-PHOTON LINKAGE

I am also interested in how the photons constitute light (the way – as a simplistic analogy – vehicles constitute a convoy.) I believe there is some type of a linkage in the sense of classical physics, but I have no specific ideas in this regard thus far. For now just accept this proposition which should not be found too objectionable: Whatever the photon-makeup of light is, it must always hold for light to remain light. If adjacent vehicles in a convoy move a mile apart, then it is no longer a convoy.

LINK FAILURE

When the distance R equals R_v, a distance so large that

$$u_v \, \Delta v \, V \sim h \, v, \qquad\qquad\qquad\qquad (D\text{-}3)$$

i.e., the stellar energy available in the volume occupied by the photon is the same as the photon energy hv, a critical juncture is reached. At greater distances there is not enough energy available for photons to exist in a connected continuum. The light-photon linkage fails when the intensity falls below a threshold.

PHOTON BANDWIDTH Δv

The bandwidth Δv which we are forced to introduce above is in fact exactly what is called for from any view of the photon as a free-standing magnetic structure. Consider (in principle) translating such an isolated structure at the velocity c past a light detector connected to a suitably fast oscilloscope. Then an electromagnetic pulse will be recorded – a pulse which begins at zero amplitude and ends in zero amplitude. When this time-domain pulse is fed to a spectrum analyzer or is digitally transformed, one obtains its corresponding frequency spectrum. This gives the dominant frequency v and the bandwidth Δv.

WHAT HAPPENS NOW?

Driven by the light-photon linkage "mechanism" the light spontaneously moves to lower frequencies and maintains the condition (D-3). I cannot see that this is forbidden in Electromagnetic Theory. This process need not violate any conservation laws – when integrated over a large volume.

And at any rate, in-flight stretching of the wavelength of light is today a fully accepted physics mechanism in Big Bang cosmology – although that occurs due to stretching of space. Why can it not occur due to some more amenable physics reason?

So past R_v, light from a star begins to redden, and any observer at a distance greater than this will see this "reddening" of light. Such reddening clearly would be detectable only with reference to spectral lines of known original frequencies.

EPILOGUE

On 30 January 2015 the European Space Agency (ESA) issued a Press Release [1] on the joint BICEP2-Planck study of the possible discovery of gravitational waves by the BICEP2 telescope (Chapter II-6), first reported on 17 March 2014.

In the face of all the stark and fatal defects of BICEP2 instrument (Chapter IV-6), ESA Planck Collaboration provided a pointedly strong endorsement of the BICEP2 technology:

So, the Planck and BICEP2 teams joined forces, combining the satellite's ability to deal with foregrounds using observations at several frequencies – including those where dust emission is strongest – with the greater sensitivity of the ground-based experiments over limited areas of the sky, thanks to their more recent, improved technology.

A joint BICEP2-Planck paper was accepted by *Physical Review Letters*. It retracted the BICEP2 discovery that was published here in June 2014. The new paper would roll back the discovery of gravitational waves but roll out further the detection of B-mode polarization swirls on the sky, with even greater confidence.

It was clear from various reports that they were going full speed ahead with BICEP2, BICEP3, Keck Array, and the SPIDER telescopes without any modifications. All these telescopes share the same defective imaging plane technology.

Thus in fact the Europeans were helping perpetuate the search for B-mode swirls by the Americans with an entire family of bogus telescopes. This new campaign was not unlike the earlier one where the former deep-sixed their own finding of the CMB spectrum that disproved the American discovery of the COBE blackbody. Big Bang was now being formally protected before the world by a NATO-like trans-Atlantic alliance.

There was no visible concern in the scientific establishment or the funding agencies that there was anything the matter with the BICEP2 instrumentation. On the contrary, a decision was made to launch the balloon-borne SPIDER on New Year's Day 2015 with the botched instrumentation on board. In a comment to space.com on 12 February 2015, BICEP2 team leader John Kovac of Harvard University said [2]:

We're in a very good place right now to make good progress...

There was a comment from BICEP2 team leader James Bock of California Institute of Technology on 2 Match 2015 [3]:

The BICEP2/Keck maps are also the best ever made, with enough sensitivity to detect signals that are a tiny fraction of the total.

There was no comment of any kind from any quarters on the COBE blackbody fraud or on suppression of crucial scientific data by WMAP and Planck satellites. NASA was in evidence maintaining the COBE discovery. John Mather was flying high.

I started this book with talk of the plight of our modern scientific civilization vis-a-vis the decline and fall of the Roman Empire. I promised to return to that point. Now I do so. What we are dealing with here is the onset of a long-term decline and fall of our scientific civilization. It is being recklessly liquidated for golden calf-like idolization and quick narcissistic fix for a few. No invading Ostrogoths or Visigoths: It is the cleverest of clever scientists taking our civilization under, from within.

10 March 2015

REFERENCES:

[1] http://www.esa.int/Our_Activities/Space_Science/Planck/Planck_gravitational_waves_remain_elusive
[2] http://www.space.com/28516-cosmic-inflation-gravitational-waves-hunt.
[3] http://www.preposterousuniverse.com/blog/2015/03/02/guest-post-an-interview-with-jamie-bock-of-bicep2/

NOTES AND REFERENCES

BOOK I

This section is based largely on standard textbook material on Big Bang cosmology. A convenient place to read further about Big Bang Cosmology is Wikipedia. It contains general articles as well as more in-depth articles on subtopics.

§I-9: With regard to dissidence three examples may be of interest.

First, in 1990 Hannes Alfvén and Carl-Gunne Fälthammar wrote a report *Can the Big Bang Survive in the Space Age?* which would be the last act of dissidence by Alfven before retiring from scientific life. The report's conclusions:

In the present letter we have, largely on the basis of experience gained from space research, discussed three different kinds of objection to the Big Bang theory.

(1) The idealisation involved in the Big Bang theory represents a departure from reliable empirical knowledge that is many powers of ten worse than that which proved gravely misleading in the history of space research.

(2) Like some elegant but unfortunately misleading theories in the history of space research, the Big Bang may explain certain observations, but this is no guarantee that it is the correct explanation.

(3) Some incontrovertible observational facts are incompatible with the Big Bang theory.

In the light of this it is easy to share Maddox' (1989) expectation that the Big Bang cannot survive long in the space-age, and to feel concern that it may play a role in cosmology all too similar to that played by the theories that delayed progress in space science for alarmingly similar reasons.

Full article at: *http://kth.diva-portal.org/smash/get/diva2:514256/FULLTEXT01.pdf*

The Maddox reference concerns a commentary published in the journal Nature by its then editor John Maddox (John Maddox, "Down with the Big Bang", *Nature,* v.340, August 10, 1989.)

In 2004 there was published *An Open Letter to the Scientific Community* in the magazine *New Scientist* (May 22-28 issue, 2004, p. 20.) The effort was led by Eric J. Lerner. An excerpt:

Today, virtually all financial and experimental resources in cosmology are devoted to big bang studies. Funding comes from only a few sources, and all the peer-review committees that control them are dominated by supporters of the big bang. As a result, the dominance of the big bang within the field has become self-sustaining, irrespective of the scientific validity of the theory.

Giving support only to projects within the big bang framework undermines a fundamental element of the scientific method -- the constant testing of theory against observation. Such a restriction makes unbiased discussion and research impossible. To redress this, we urge those agencies that fund work in cosmology to set aside a significant fraction of their funding for investigations into alternative theories and observational contradictions of the big bang. To avoid bias, the peer review committee that allocates such funds could be composed of astronomers and physicists from outside the field of cosmology.

The signatories to this letter included Halton C. Arp, Hermann Bondi, Thomas Gold and Jayant Narlikar.

In 2004 a nine-part video series titled *The Big Bang Never Happened* was produced. It includes commentaries from Halton C. Arp, Jeffrey and Margaret Burbidge, Fred Hoyle, Eric Lerner, Jayant Narlikar, and others.

Web site:
http://www.worldsci.org/php/index.php?tab0=More&tab1=Media&tab2=Display&id=222

This video series takes its name from the book *The Big Bang Never Happened* by Eric J. Lerner (Vintage Books, 1992).

BOOK II

§II-1: The Penzias-Wilson and Dicke et al back-to-back papers:

R. H. Dicke, P. J. E. Peebles, P. G. Roll, and D. T. Wilkinson, Astrophysical Journal, vol. 142, pp. 414-419 (1965).

A. A. Penzias and R. W. Wilson, Astrophysical Journal, vol. 142, pp. 419-421 (1965).

§II-2.2: Results reported by the Richards group:

D. P. Woody et al, Measurement of the Spectrum of the Submillimeter Cosmic Background, Physical Review Letters, vo. 34, no. 16, 1036-1039, 1975.

D. P. Woody and P. L. Richards, Near-millimeter spectrum of the microwave background, Astrophysical Journal, 248: 18-37, 1981.

Representative scientific papers on Mather COBE satellite discovery:

J. C. Mather, D. J. Fixsen, and R. A. Shafer, "Design for the COBE far-infrared absolute spectrophotometer", Proc. SPIE2019, Infrared Spaceborne Remote Sensing, 168 (October 1, 1993).

J. C. Mather, Nobel Lecture: From Big Bang to the Nobel Prize and beyond, Reviews of Modern Physics, vol. 79, October-December 2007.

For general interest reading on the John Mather COBE satellite experiment, see the following book:

John Mather and John Boslough, The Very First Light: The True Inside Story of the Scientific Journey Back to the Dawn of the Universe (Basic Books, 2008).

§II-3: Representative scientific papers on Smoot discovery:

C. L. Bennett et al, COBE differential microwave radiometer: Calibration techniques, Astrophysical Journal, vol. 391, pp. 466-482 (1992).

For general interest reading on the George Smoot COBE satellite experiment, see the following book:

George Smoot and Keay Davidson, Wrinkles in Time: Witness to the Birth of the Universe (Harper Perennial, 2007).

§II-3 Representative scientific papers on Gush discovery:

H. P. Gush, M. Halpern, and E. H. Wishnow, Rocket Measurement of the Cosmic-Background-Radiation mm-Wave Spectrum, Physical Review Letters, vol. 65, no. 5, (30 July 1990).

§II-4 A general account of the Perlmutter discovery:

Saul Perlmutter, Supernovae, Dark Energy, and the Acceleration Universe, Physics Today (April 2003).

More references are found there.

CHAPTER II-6 For a description of the BICEP2 project, see:

P. A. R. Ade et al. , Detection of B-Mode Polarization at Degree Angular Scales by BICEP2, Phys. Rev. Lett. 112, 241101 – Published 19 June 2014

§II-6.2 This section is largely based on the following report dated 29 October 2014:

From Discovery to Dust by Amanda Gefter
http://www.pbs.org/wgbh/nova/next/physics/bicep2/

BOOK III

Book III is based largely on textbook material one can readily find. Some material here may be experience-based and practice-based know-how, not explicitly written down in textbooks.

BOOK IV

§IV-2.2 John C. Mather Ph. D. Thesis, University of California, Berkeley (1974):

Far infrared spectrometry of the cosmic background radiation
http://www.osti.gov/accomplishments/documents/fullText/ACC0 461.pdf

§IV-2.7 The pilot study report:

J. C. Mather, M. Toral, and H. Hemmati, "Heat trap with flare as multimode antenna", Applied Optics, vol. 25, issue 16, pp. 2826-2830 (1986).

§IV-2.12 On the calibration technique:

J. C. Mather et al, Calibrator design for the COBE Far Infrared Absolute Spectrophotometer (FIRAS), Astrophysical Journal, 512:511 –520, 1999.

§IV-2.13 The paper on adding the flare section to the Winston Cone:

J. C. Mather, "Broad-band flared horn with low sidelobes", IEEE Transactions on Antennas and Propagation, Vol. 29, Issue 6, pp. 967-969 (1981).

§IV-2.14 For FIRAS orbital failure, see:

J. C. Mather, D. J. Fixsen, and R. A. Shafer, Design for the far-infrared absolute spectrophotometer, Proc. SPIE 2019, Infrared Spaceborne Remote Sensing, 168 (October 1, 1993).

§IV-2.16 The photo of the actual FIRAS antenna is published here:

http://lambda.gsfc.nasa.gov/product/cobe/about_firas.cfm

§IV-2.17: The COBE moon sweep diagram is in the paper by Mather, Fixsen and Shafer (1993); see full reference under §IV-2.14.

§IV-2.21: The video is here:

John Mather Maniac Lecture 2013

https://www.youtube.com/watch?v=wuByPMRU0ME

§IV-3.5 Smoot's instruments are described here:

G. Smoot et al, COBE differential microwave radiometers: Instrument design and implementation, Astrophysical Journal, 360: 685-695 (1990)

§IV-3.7: Smoot's moon sweep figures are here:

C. L. Bennett et al, COBE differential microwave radiometers: Calibration techniques, Astrophysical Journal, 391: 466-482 (1992).

§IV-3.8 The quotation is from the above reference, page 476.

§IV-3.10 *The quotation is from:*

C. Barnes et al, First-year Wilkinson Microwave Anisotropy Probe (WMAP) 1. Observations: Galactic signal contamination from sidelobe pickup, Astrophysical journal Supplement Series, 148:51–62 (2003).

WMAP moon sweep is here:

R. S. Hill et al, Five-Year Wilkinson Microwave Anisotropy Probe (WMAP1) Observations: Beam Maps and Window Functions, Astrophysical Journal Supplement Series, 180:246-264 (2009).

§IV-5.3 On White Mountain research:

G. F. Smoot, The Spectrum of the Cosmic Microwave Background Radiation: Early and recent Measurements from the White Mountain Research Station by George F. Smoot

NASA STI/Recon Technical Report N 09/1985.

§IV-6.7 The Roger O'Brient quotation is taken from:

http://www.astro.caltech.edu/~rogero/detectors.html

§IV-6.8 James Bock Quotation about telescope:

BUILDING BICEP2: A conversation with Jamie Bock by Cynthia Eller 3/17/14

http://www.caltech.edu/news/building-bicep2-conversation-jamie-bock-42306

§IV-7 CMB angular power spectrum:

Cosmic microwave background

http://en.wikipedia.org/wiki/Cosmic_microwave_background

§IV-7 Paul Steinhardt commentary:

Big Bang blunder bursts the multiverse bubble, Nature 3 June 2014

http://www.nature.com/news/big-bang-blunder-bursts-the-multiverse-bubble-1.15346

§V-2.2 John Mather Commencement speech:
University of Maryland Winter Commencement Address, December 2008
http://www.c-span.org/video/?282999-1/university-maryland-winter-commencement-address

The Nobel YouTube exchange:
Answer: Has anyone contradicted your theory?
https://www.youtube.com/watch?v=1Th0U-C3STY&list=PLF1A02E39EC712052&index=10

§V-2.4 The Maniac Lecture 2014:
John Mather Maniac Lecture, November 19, 2014
https://www.youtube.com/watch?v=wuByPMRU0ME

§V-2.5: Smithsonian Lecture:
Behind the Science with Joe Palca: Insights from Scientific Innovators, Evening Lecture
https://smithsonianassociates.org/ticketing/tickets/reserve.aspx?performanceNumber=230110

§V-6.4 The Nobel Scholar Program:
John Mather Nobel Scholars
https://www.spacegrant.org/mather

§V-6.5 Albert Einstein World Award for Science:
http://www.consejoculturalmundial.org/winners-science.php

AIP Mather Policy Intern program:
Host: American Institute of Physics and Congressional Offices on Capitol Hill (2 available)
http://www.spsnational.org/programs/internships/positions/

§V-7.1 The Weinberg quotation:
The ever-smaller future of physics
http://news.harvard.edu/gazette/story/2014/12/the-ever-smaller-future-of-physics/

The Hawking quotation:
EXCLUSIVE: WIRED meets Professor Stephen Hawking
*http://www.wired.co.uk/magazine/archive/2015/01/features/wire
d-meets-professor-hawking/page/2*

§V-7.2 The quotations from P. J. E. Peebles are taken from:
A Cosmologist's intellectual journey
http://globetrotter.berkeley.edu/people6/Peebles/peebles-con6.html
Interview with Dr. P. J. E. Peebles
http://www.aip.org/history/ohilist/25507_1.html

§V-7.3: David Gross NOVA quotation:
Viewpoints on String Theory
David Gross
http://www.pbs.org/wgbh/nova/elegant/view-gross.html

David Gross Lindau Nobel Meetings quotation:
Betting on the cosmos - with David Gross and Robert Laughlin
https://www.youtube.com/watch?v=beQ9fZ0jVdE

§V-7.4 The Decadal Survey:
ASTRO2010: Charting the Next Decade in Astronomy
*http://www.kavlifoundation.org/science-spotlights/charting-next-
decade-astronomy#.VKHLvV2AA*

Roger Blandford testimony:
*The next great observatory: Assessing the James Webb Space
Telescope*
*http://www.gpo.gov/fdsys/pkg/CHRG-
112hhrg72165/html/CHRG-112hhrg72165.htm*

§V-7.8 The Amber Miller reference:
Ripples from the Big Bang: Listening to the beginning of time
*http://www.worldsciencefestival.com/programs/ripples_from_the
_big_bang/*

The Mario Livio reference:
Commentary: BICEP2's B modes: Big Bang or dust?
*http://scitation.aip.org/content/aip/magazine/physicstoday/article
/67/12/10.1063/PT.3.2598*

§V-7.11 John Mather's Notre Dame honorary degree citation:

Mather receives Honorary Degree

https://physics.nd.edu/news/22071-mather-receives-honorary-degree-as-department-confers-32-undergraduate-degrees/

§V-10.2 The Kohlenberger matter:

Commercial Space Backer Leaving White House Post

http://spacenews.com/commercial-space-backer-leaving-white-house-post/

§V-14 The two Andrei Linde quotations are taken from:

Stanford Professor Andrei Linde celebrates physics breakthrough

https://www.youtube.com/watch?v=ZlfIVEy_YOA

§V-13The quotation from the three Popes on page 349 are taken from the following source:

Pope Pius XII – The proofs for the existence of God in the light of modern natural science (1951)

http://www.ewtn.com/library/PAPALDOC/P12EXIST.HTM

Pope Benedict XVI (writing in 2004 as Cardinal Ratzinger) and Pope Francis (2014):

Catholic Church and Evolution

http://en.wikipedia.org/wiki/Catholic_Church_and_evolution

APPENDIX A

Alan Stern quote:

NASA News Release April 2, 2007

http://www.nasa.gov/home/hqnews/2007/apr/HQ_07080_Science_MovesIntern.html

John Mather Nobel Biography

http://www.nobelprize.org/nobel_prizes/physics/laureates/2006/mather-bio.html

Per Carlson quotation:

2 in U.S. Win Nobel Prize for Research of Universe's Origin

October 4, 2006
  http://www.washingtonpost.com/wp-dyn/content/article/2006/10/03/AR2006100300098.html
  Alan Stern news:
  Stern Steps Down as NASA Science Chief After Mars Budget Dustup
  http://www.scientificamerican.com/article/stern-steps-down-as-head/
  Bob Park report on John Mather resignation:
  What's New by Bob Park
  NASA: Resignation of two top scientists unrelated.
  Friday, March 28, 2008
  http://bobpark.physics.umd.edu/WN08/wn032808.html
  Alan Stern quotation:
  Wielding a Cost-Cutting Ax, and Often, at NASA
  http://www.nytimes.com/2008/01/01/science/space/01stern.html

APPENDIX B

NOVA: Hunting the Edge of Space
  https://www.youtube.com/watch?v=O78lISatzRI
  Transcript of above:
  http://www.pbs.org/wgbh/nova/space/hunting-edge-space.html#hunting-edge-space-2

APPENDIX D

Link to my photon paper:
  http://bibhasde.com/photon.pdf

EDUCATION

Bibhas De received his Bachelor's degree with Honors in Physics from Presidency College, University of Calcutta. He studied for a year at the Institute of Radiophysics and Electronics at the same university before coming to the United States with scholarships from UNESCO and the International Astronomical Union. He attended the University of Michigan, Ann Arbor, where he received his Master's degree in Astronomy. Here he gained training in radio astronomical observation and instrumentation techniques under the guidance of the renowned radio astronomer Fred T. Haddock and his associates. De was assigned diverse instructive projects (see e.g., [1]). He was an observer of linear polarization of variable flux radio sources (using a rotating horn polarimeter), and also searched for the H91α radio recombination line in the solar corona (using a 22-channel spectral line receiver.) He spent two summers as an observatory technician. He assisted in analyzing solar radio burst data from OGO-V satellite. He later received his Ph. D. in Applied Physics from the University of California, San Diego in 1973 as a student of Hannes Alfvén.

RESEARCH AND DEVELOPMENT BACKGROUND

Bibhas De has published contributions to the foundation of electromagnetic theory [2, 3]. His contribution to astrophysics is exemplified by references [4, 5], and to space physics by reference [6]. He has worked in the field of antenna and microwave engineering in the satellite communication industry. He received hands-on training in antenna range measurements at the very source, Scientific-Atlanta Inc., a leader in the field. Here he was responsible for the antenna engineering task, for example, in the LANDSAT IV (with dual band satellite-tracking feed horn arrays) and the INTELSAT IVA (with corrugated horn transmit/receive feed) satellite Earth Stations. Later, while in the oil industry, he engaged in extensive radiofrequency and optical laboratory experimentations, in conjunction with related theory development [7, 8]. He supervised a group of laboratories engaged in precision measurement of petrophysical and electrical properties of petroleum reservoir rock samples. He also developed methods of measurement of dielectric properties of subsurface rock formations by instruments lowered deep in a drill hole, and also in other applications. He holds a number of patents in these areas [9, 10].

NASA background disclosure: Bibhas De's graduate studies and a two-year postdoctoral appointment were supported by NASA grants to the University of California, San Diego. Later he spent three years at the Lunar and Planetary Institute, Houston, supported by NASA.

REFERENCES

1. B. R. De, A survey of manmade radio signals between 4 GHz and 11.5 GHz, University of Michigan Radioastronomy Observatory Report 70-7.

2. B. R. De, Gravitational mass of magnetostatic field. *Astrophysics and Space Science* **239**, 25-33, 1996.

3. B. R. De, Magnetohydroelectric waves in a fluid dielectric. *Physics of Fluids* **22(1)**, 189-190, 1979.

4. B. R. De, On the Mechanism of formation of loop prominences. *Solar Physics* **31**, 437-447, 1973.

5. B. R. De, Disequilibrium condensation environments in space: A frontier in thermodynamics. *Astrophysics and Space Science* **65**, 191-198, 1979.

6. B. R. De, Conductor-like behavior of photo-emitting dielectric surfaces, *Journal of Geophysical Research - Space Physics*, **84**, 2655-2656, 1979.

7. D. Stroud, G. W. Milton, and B. R. De, Analytical model for the dielectric response of brine-saturated rocks. *Physical Review* **B 34**, no. 8-I, 5145-5153, 1986.

8. B. R. De, I. H. H. Hoffmann, D. Stroud, and M. A. Nelson, Optical analog of the permeability of sandstones. *Physical Review* **B 45**, 196-204, 1992.

9. B. R. De (inventor), Sample accommodator and method for the measurement of dielectric properties. United States Patent 4,866,371 issued Sep. 12, 1989.

10. B. R. De and M. A. Nelson (inventors), Method and apparatus for combined dielectric and resistivity oil well logging over a wide range of frequencies. United States Patent 5,132,623 issued July 21, 1992.

INDEX OF NAMES

www.ingramcontent.com/pod-product-compliance
Lightning Source LLC
Chambersburg PA
CBHW021419170526
45164CB00001B/9